EVERYDAY LIFE

IN THE

NEW STONE, BRONZE
& EARLY IRON AGES

British Library Cataloguing-in-Publication Data
A catalogue record for this book is available from
the British Library

THE EVERYDAY LIFE SERIES

By MARJORIE and C. H. B. QUENNELL

Now completed in Four Volumes. Full Illustrated Prospectus on application.

I.—EVERYDAY LIFE IN THE OLD STONE AGE

Containing 128 pages, including 70 Illustrations and a Coloured Frontispiece, from Drawings specially prepared by the Authors. Third Edition revised. Large Crown 8vo. (7″ × 5½″), cloth, 7s. 6d. net.

" With this book as a guide, those for whom it is written will be able to begin their prehistoric studies under the pleasantest auspices and, it may be hoped, will be inspired to go still further."
The Antiquaries' Journal.

II.—EVERYDAY LIFE IN THE NEW STONE, BRONZE AND EARLY IRON AGES.

Containing 136 pages, including 90 Illustrations and 2 Coloured Plates. Third Edition. Large Crown 8vo, cloth, 7s. 6d. net.

" When the writers are the Quennells it seems almost superfluous to add that it is a very attractive and vividly written book, accurate without being burdened with technicalities, and full of delightful illustrations both in line and colour."—*G. K.'s Weekly.*

Parts I. and II. of "The Everyday Life Series" may now be purchased complete in one volume. Crown 8vo, cloth, 15s. net.

EVERYDAY LIFE IN PREHISTORIC TIMES.

Containing 223 pages, including 160 Illustrations and 3 Coloured Plates.

" A new ' Quennell book ' is an educational event. For children or for adults it would be difficult to find a better short account of early man than this volume offers."—*The Education Outlook.*

III.—EVERYDAY LIFE IN ROMAN BRITAIN.

Containing 108 pages, including 101 Illustrations and 3 Coloured Plates. Second Edition. Large Crown 8vo, cloth, 7s. 6d. net.

" Roman Britain becomes something real before the reader's eyes: and the pictures are as good as the text."—*The Daily Mail.*

IV.—EVERYDAY LIFE IN NORMAN, VIKING AND SAXON TIMES

Containing 128 pages, including 80 Illustrations and 2 Coloured Plates. Second Edition. Large Crown 8vo, cloth, 7s. 6d. net.

" It is a period which gives scope for interesting writing and delightful illustrations. The authors have, as before, profited to the full by their opportunities. The illustrations are well chosen. The most charming are, however, those of ecclesiastical architecture. Altogether this is an agreeable as well as a valuable book, and one can say of the authors what Asser said of Alfred. They are ' affable and pleasant to all, and curiously eager to investigate things unknown.' "—*The Times.*

FIG. 1.—Warriors and Chariot of the Early Iron Age.

To face Title-page.

THE EVERYDAY LIFE SERIES—II

EVERYDAY LIFE

IN THE

NEW STONE, BRONZE
& EARLY IRON AGES

WRITTEN & ILLUSTRATED BY
MARJORIE & C. H. B. QUENNELL
Authors of "Everyday Things in England"

THIRD EDITION, REVISED

CONTENTS

		PAGE
INTRODUCTION TO THIRD EDITION	. .	vi
SHORT LIST OF AUTHORITIES	. . .	x
MAP OF BRITISH ROAD SYSTEM, ETC.	.	xi
CHART	. . . . . *To face*	1

CHAPTER

I. THE NEW STONE AGE . . . 1

Before the New Stone Age—Kitchen Middens—Migrations—Geographical Conditions—The Naked Chalk—Trackways—Camps—Iberians—Celtic and Nordic Men Aryan-speaking People — Flint Implements — Cores—Flakes—Axes—Arrows—Huts—Hut Circles—Fires—Cooking—Corn—Cakes—Pots—Pans—Earthworks—Fortification—Gateways—Water—Life in the Wild—Trapping—Civil Engineering—Long Barrows—Tombs—Houses—Towers—Rough Stone Monuments—Leverage—Wedges—Building Stonehenge—Sun Temples—Nature Worship.

II. THE BRONZE AGE 51

Bronze-Smelting—Swords and Spears—Heathery Burn—Spinning — Looms — Weaving — Costume — Razors — Wheels — Communications — Harvest — Pottery — Burial—Barrows—Hector—Patroklos—Trade—Trade Routes—Migrations—Tin—Trackways—Conditions of Life.

III. THE EARLY IRON AGE . . . 82

Lake Dwellings—Glastonbury—Huts—St. Paul's Cathedral—David Cox—Ploughs—Smelting—Knives—Tools—Brooches—Lathes—The Axe—Iberians—Boats—Spears—Enamels — Chariots — Burials — Trackways — Surveying—Currency Bars—Conditions of Life—Celtic Legends.

INDEX TO TEXT AND ILLUSTRATIONS 118

FIG. 2.—The Linces, Cheddington, Bucks.

INTRODUCTION TO THIRD EDITION

SINCE Mr. and Mrs. Quennell prepared this book, and revised the second edition of it, an unprecedented amount of research by excavation into the intriguing mysteries of Prehistory has been done. H.M. Ministry of Works, learned societies, and keen individuals have gone to work in all parts of the country, delving and comparing notes to an extent never dreamed of by the earlier schools of archæology. In addition, older methods of divination for buried treasure were reinforced by the aeroplane. It was found that man, given a **real** bird's-eye view of things, could not be seen on the ground itself ; and the two most interesting surprises were sprung on us, not by the antiquaries but by a V.C. air-pilot of the then young R.A.F.

The finds and results of the great quest have, in the main, gone to confirm most of our earlier theories and to add a great deal of knowledge where gaps previously existed. And that is surely satisfactory when one recalls how often in the past fresh exploration has done nearly as much to overthrow beliefs which had been held as to reveal new truths.

So far as doubts are raised by the new research I think that the only comment which can be made on Mr. Quennell's account is a warning against too strict an adherence to formulated rules of long heads and round heads, though the old tag about long heads in long barrows and round heads in round barrows is still true in the main. There have been too many contradictions in craniology to allow us to dogmatise yet about racial extraction. Also the older belief that the Goidels (" Q " Celts) ancestors of the Irish, Highland, and Manx stocks, were solely responsible for the earlier Iron Age immigrations, while the Brythons (" P " Celts) forebears of the Welsh and Bretons brought in the later Iron Age, is no longer held firmly. The only thing that seems certain is that all the

INTRODUCTION TO THIRD EDITION

principal clans involved in the Iron Age invasions were Celtic
and that Celts were in the ascendancy in Britain when the
Romans arrived ; and hints as to how the two branches were
distributed can still be gleaned from the names of rivers (so
fluid, yet so conservative, resisting four linguistic invasions).
Avons, Derwents, Doves, Calders, and Dees indicate Brythonic
(Welsh) settlement. They are to be found in many parts of
England and Scotland as well as Wales. Exe, Axe, Usk, and
Esk, is the sign of the Goidel. Outside Ireland this intruder
seems to have confined himself to the Dartmoor watershed
and the western seaboard.

Among the greater enterprises of recent years was the excavation and partial renovation of the huge circle at Avebury,
the remains of whose vast rampart and cyclopean colonnade
enclosed the greater part of a village. Here, and in the equally
imposing Kennet alignment, adjacent, stones which had been
thrown down were re-erected and a section of the great ditch
cleared out. Not a great deal of new knowledge accrued from
this extensive work. The mystery of the cult of the stone
circle remained as obscure as ever. But finds demonstrated
that Avebury and therefore, assuredly, Stonehenge must be
brought forward from the Neolithic Period into the dawn of
Bronze.

It was about this time that Wing-Commander Insull made
the first of his two really exciting discoveries, flying over a
cornfield at Durrington, two miles from Stonehenge, at a
height of two thousand feet. Below him he saw the ghost-like
image of six concentric circles of dotted rings. The dots, on
investigation proved to be empty sockets of the colonnades of a
prehistoric temple. Filled up with the dust of ages, they were
invisible from ground level, but being of a different density
from the undisturbed native loam, they caused just that slight
difference in the growth of the grain to make them discernible
from a height. It was demonstrated, however, that these
sockets had not held stone pillars but wooden ones. It is
thought that here was the prototype and forerunner of Stonehenge and the location has been given a name new to geography—Woodhenge (Fig. 44). Concrete stumps now mark
the post-holes.

Those who had paralleled Stonehenge with the sun-dance
lodges of the Redskins had said that it must have had wooden
predecessors though it was never expected that any trace
could remain. Wing-Commander Insull looked again and
found another Woodhenge, this time not on Salisbury Plain
but in that of Norfolk, at Arminghall, near Norwich. It was

INTRODUCTION TO THIRD EDITION

really more like Stonehenge than the other, for the colonnade (though a single one only) was horseshoe shaped like that of the great trilithons.

One other monument should be mentioned as falling within the early Bronze Age, though it has all the massiveness of the Neolithic builder. This is the tumulus at Bryn Celli Dhu in Anglesey. It is a large round barrow with a passage leading into the heart of it, where is found a single chamber, in which stands a smooth-tooled pillar of stone nearly the height of a man (Figs. 28, 29, pp. 31, 32). Excavation here revealed a number of new features in prehistoric ritual of the greatest interest, but of too complex a nature to mention in this brief summary. The lay-out and construction of this tumulus would seem to indicate the fusion of many cults, both of the tomb and of the temple-circle. But still we are kept guessing, uninitiated even into the first degree of the mystery. One of the riddles posed here, though, I cannot resist illustrating, as it is one of the earliest attempts at picture-writing in our country—a voice crying in a wilderness of stone. The hieroglyph in Fig. 29a (p. 31) was made on three sides of a rectangular block of stone buried flat, like a lid on a box, near the midst of the mound, but outside the chamber. Below it was laid a lump of red jasper.

While these new finds add to our wonder rather than to a clarification of the past, it is otherwise with the work that has been done on the hill-forts since Mr. Quennell described them on pp. 6 and 28. It seems certain that *forts* (as they are now more generally styled, rather than *camps*) were built by the Neolithic people but not by those of the succeeding Bronze Age. There is, so far, no evidence to show that any fort-making took place during a lapse of perhaps 1500 years, the great monuments of the Bronze Agers being those of religion and not war, a striking fact indeed ! When the first Iron Age immigrants appeared, about 500 B.C. the fort-building restarted, sometimes on the ancient Stone Age sites which were improved, sometimes on new ground. Successive waves introduced new and more up-to-date defensive measures until the Romans with a more efficient simplification of the whole art of war cut short the evolution of the earthwork fort, which was becoming frightful in its cumbersomeness and complexity.

At Windmill Hill, in Wiltshire, the simple fort of the Stone Ager is seen with all the finds housed conveniently close by in the museum at Avebury. The outstanding feature (not yet explained), is that the ditch is not carried in its entirety (like a moat) right round the fort, but interrupted at frequent regular intervals by causeways. It looks almost as if the place were indeed a genuine *camp* for man and beast, the ditch being merely a quarry to afford material for raising the rampart, which was to keep animals from getting out and not men from getting in. In 1934 and 1935 Maiden Castle in Dorset (p. 28) was excavated and the whole story of successive occupations of the hilltop was laid bare, from the time when Neolithic man cut his causeway-ditch to compass only half the present site to the last days of pre-Roman Britain, when some great chief of a Belgic clan made the final contribution to its complex defences. Though the excavations were of small extent, considering the area of the site, not only was the military history of Maiden Castle revealed in epitome, but that of domestic conditions as well, for it was seen how for centuries a large population had lived there in a permanent town, and had been driven through shortage of space to make dwellings on sites which even then must have been considered both undesirable and insanitary. Dr. Mortimer Wheeler's book on this excavation is one of the most entrancing of its kind ever brought out.

At Ladle Hill, in Hampshire, there is to be seen a fort in course of construction, abandoned at an elementary stage of its making for some reason lost to history, and above Llanidloes in Mid-Wales is one whose completion was interrupted in the final stage. More interesting still, though, is that some forts which threatened Roman supremacy were *slighted*, to borrow an expression usually applied only to mediæval castles, signifying that a fortress has been rendered indefensible by having its rampart breached and cast down into the ditch. Evidence that this was done is forthcoming at Eddisbury in Cheshire and Almondbury in the Pennines.

In the chronological chart on p. xi the more recent academic divisions of the Ages of Bronze and Iron have been substituted for those given in the First and Second Editions. They should not of course be thought of as more than temporary parcellings out of time and culture which will be subject to many a modification yet as the knowledge of our Ancient British ancestry increases.

E. V.

Winter, 1945.

CHART

BIRTH OF CHRIST

EGYPT	GREECE	ASSYRIA & BABYLON	ROME	BRITAIN

EGYPT

Dynasties:
Becomes a province of the Roman Empire on death of Cleopatra, 30 B.C.

Ptolemies, 304. Built lighthouse, Pharos. Alexander the Great, 332.
30th & 31st, 358. Persians again conquer for eight years.
29th, 399.
28th, 420. Overthrow of Persians.
27th, 527. Persian kings expelled.
25th, 700. Assyrians conquer Egypt, 670.
24th.
23rd, 766.
22nd, 950. Capture of Jerusalem.

21st, about 1050. Divided into two kingdoms.
20th.

Exodus of Israelites.

19th, B.C. 1370.

Karnak, Luxor, and 'Cleopatra's Needle.'
18th, B.C. 1600. The New Empire.
17th }
16th } Shepherds — Semitic nomads from the E.
15th }

14th. The succession of kings from the 13th to the 18th Dynasties is unknown.
13th.

The rule of this Dynasty was prosperous, and the arts flourished.

12th, B.C. 2466. Middle Empire.
11th, B.C. 2600.

Of the 7th to the 10th Dynasties little is known, and the period appears to have been one of disorder.

6th, B.C. 3300.

Worship of Sun-god.

5th, B.C. 3566. and & 3rd Pyramids built. Cheops builds Great Pyramid.
4th, B.C. 3733. End of Archaic Period. Step Pyramid at Sakkárah.

3rd, B.C. 3966.

Egyptians skilled in metal working.
2nd, B.C. 4133. Khá-Sekhemui king.
Worship of Osiris began to displace the worship of Ancestors.

1st Dynasty. Mená king.

Neolithic and Palæolithic Period.

(Age column: Iron Age — Late Minoan Bronze — Middle Minoan Bronze — Early Minoan Bronze)

GREECE

CRETE.

Third Macedonian War, 168. Second, 197.
First Macedonian War, 205.
Plato begins teaching, 386.
Parthenon commenced, 447.
Birth of Herodotus, 484.
Greeks reach Marseilles.
Rise of Sparta.
1st Olympiad, 776.

Ionian migration.
Overthrow of Minoan civilization (Mediterranean & non-Aryan).
Dorian migration.
Trojan War—Homeric Age.
Sixth City of Troy.
Decline of power and destruction of Cnossus.
Rebuilding of Palace of Cnossus.
Immigration of Achæans (Aryan) from N.
Bronze Age in Greece.
Catastrophe at Cnossus.
Rise of Mycenæ.
Later palaces at Cnossus and Phæstus built.
Palaces at Cnossus and Phæstus destroyed.
First palaces at Cnossus.
Second City of Troy.
Spiral used as decoration.
Neolithic civilization.

ASSYRIA & BABYLON

Alexander defeated Darius III., 332.
Persians, 538.
Jews exiled to Babylon by Nebuchadnezzar II., 587 B.C.
Scythians invade Asia, c. 600.
Sargon II & Sennacherib, Kings of Assyria.
Tiglath-Pileser III. of the Bible, 745. Shalmaneser.
Tiglath-Pileser II. King of Assyria, 744.
and Assyrian Empire, 950–800.
Solomon, King, about 960.
David, King, about 990.
Tiglath-Pileser I., 1130–1100.

1st Babylonian Dynasty overthrown.

Arabian nomads acquire the horse.

(Age column: Iron — Late Bronze, Kassite, Babylonian & Early Assyrian — Middle Bronze, Akkadian & later Sumerian — Early Bronze, Early Sumerian)

ROME

Destruction of Carthage, 146.
First Punic War, 264.
Latin and Samnite Wars, 340.
Sack of Rome by Gauls, 395 B.C.
Foundation of Republic, 509.
Rome founded, 753.
Cadiz founded by Phœnicians, about 1100.

N.B.—The dates on this Chart have been gathered from many sources, and the authorities do not agree; there are, for example, at least six systems of chronology which have been formulated by Egyptologists; it is therefore a very difficult matter to present even a suggestion which will not be challenged, but we publish this Chart because we are anxious not to shut off our History into watertight compartments. We want our readers to bear the relation of Britain to the World in mind. If we are a few hundreds of years wrong here and there, what matter? Let them discover our mistakes and learn by them.

BRITAIN

Roman occupation of Britain, A.D. 43.
Julius Cæsar raids Britain, 55 and 54 B.C.
Iron Age C. Invasion of the Belgæ. A mixed race, partly Celtic, partly Germanic. About 75 B.C.
Iron Age B. About 250 B.C.
Iron Age A. About 500 B.C. — Celts
Late Bronze Age. About 1000 B.C.
Middle Bronze Age. About 1600 B.C.
Early Bronze Age. About 1800 B.C.
Beaker People. About 1900 B.C.

(Age column: Iron — Bronze — Neolithic)

YEAR

MAIDEN CASTLE DORSET

(Plan with labels: NEOLITHIC, LONG MOUND, ROMAN TEMPLE, SITE A, SITE B, TUMULUS, WELL, and contour heights 430, 440, etc.)

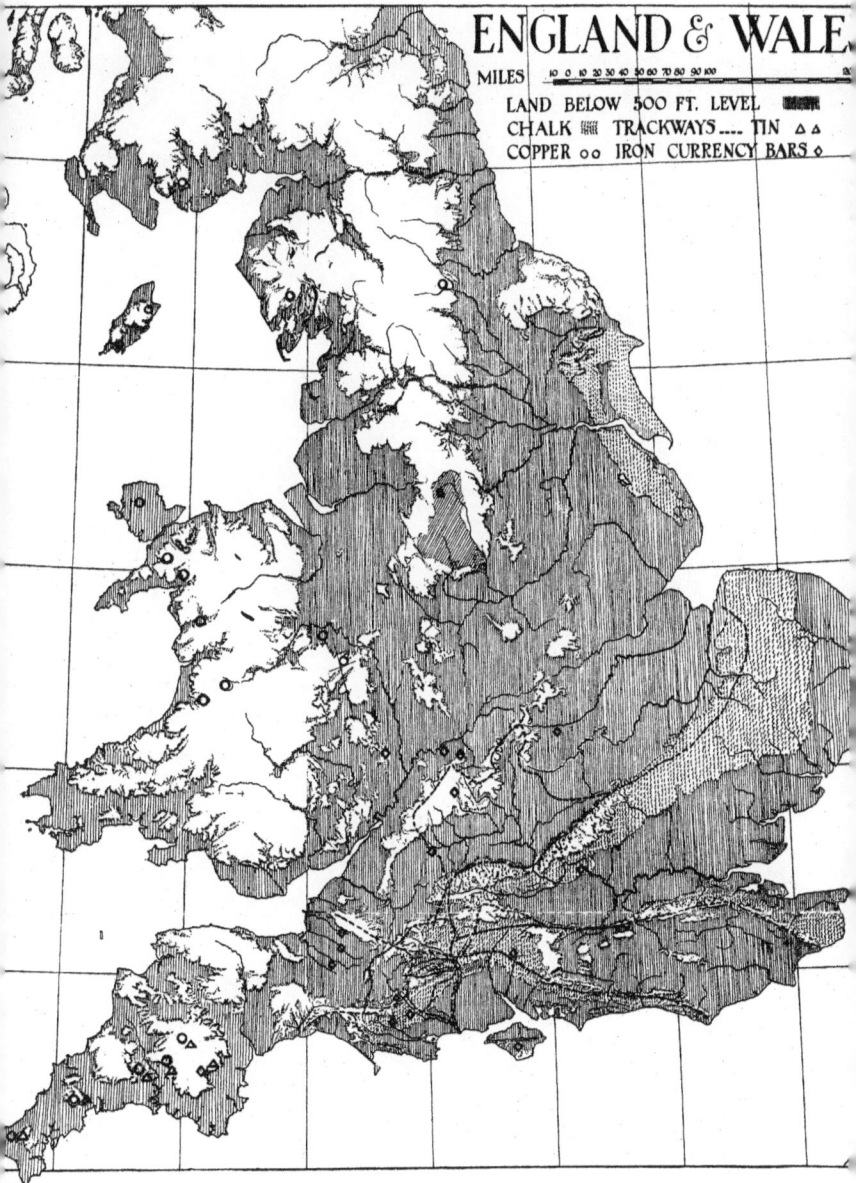

FIG. 3.—Map showing British Road System, Levels, Chalk, Minerals, and Currency Bars.

EVERYDAY LIFE IN THE NEW STONE, BRONZE & EARLY IRON AGES

CHAPTER I

THE NEW STONE AGE

BEFORE THE NEW STONE AGE

BEFORE we begin with the doings of the men of the Neolithic or New Stone Age, it may be as well to give our readers a reminder of the periods which are associated with the Palæolithic or Old Stone Age with which we dealt in Part I.

We started with the period of the River Drift, so called because of the flint implements found in the gravels deposited by rivers. Man lived on the banks of the Thames up to Oxford ; along the Lea to the Dunstable area ; around the Solent and Avon in Hampshire, and the Wey at Farnham, and on an area in E. Anglia, bounded by Thetford, Hoxne, Bury St. Edmunds, Mildenhall, and Lakenheath.

Then we came to a period when men lived in caves, like Kent's Cavern and Brixham Cave, N. and S. of Tor Bay, Wookey Hole in Somerset, Cresswell Caves in Derbyshire, and others in Wales.

Finally we saw how, at the end of the Old Stone Age, man seemed to have been drawn, or driven, to the water. The people, called Azilian, after Mas d'Azil (France), lived on great rafts anchored in the middle of lakes, as at Maglemose, Denmark. At Oban in Scotland, Azilian deposits were found in a cave opening on to a seabeach. This Azilian

KITCHEN MIDDENS

civilization is the first of which we have any evidence in Scotland during the Old Stone Age, and we must not forget that the Northern part of Great Britain was covered with ice during the Glacial periods, and probably was too bleak and desolate in the Interglacial periods to attract settlers, until the ice had finally retreated in early Neolithic times. France was always ahead of us in civilization, because the greater part of it was never glaciated.

At Oban were found the bones of large sea fish, red deer, goat, pig, and many other animals, and the life led there must have resembled that which we trace in the Kitchen Middens on the Danish coast. These middens are of the greatest interest, because they belong mostly to the earliest Neolithic period, and it is here that we shall start this, Part II. of our series.

THE SHELL-MOUND OR KITCHEN MIDDEN PEOPLE

A midden is a rubbish heap, and in Denmark these mounds are sometimes 100 yards long, by 50 wide, by 1 high, and were formed of the refuse of the meals and life of prehistoric man. They are labelled there with the splendid name of *Kjökkenmöddinger*, and are largely formed of oyster shells, with the bones of stag, roe-deer, and wild boar. The long bones have been cracked to extract the marrow. The people do not appear to have grown any crops, or domesticated any animals, except the dog, so they had not made any great advances on the civilization of the Old Stone Age. It was the pleasant loafing life of the beach-comber. The sea when it is angry casts up all kinds of edible flotsam, and in kindlier mood, at low tide, early Neolithic man could hunt over the rocks, as we do to-day during our summer holidays, and find lobster and crab, oyster and mussel, prawns and shrimps, and the humble winkle.

We find the remains of similar people, and their shell heaps, in different parts of the British Isles, and at the British Museum, in the Prehistoric Room, are flints from the Castle Hill at Hastings. These people possessed dug-out canoes, or skin-covered boats, with which to go fishing, and used harpoons like the Old Stone Age men. It may be that, as their flint implements were rough and not very effective, they were forced to the seaside by the encroaching forests. As the weather improved, after the

Ice Ages, the trees grew, and man could not as yet make sufficient clearings in which to start agriculture.

The evidence that we can gain, points to this dim beginning of the Neolithic period, some 7000 to 10,000 years ago, as a time when the world was gathering its forces. The Old Stone Age culminated in the wonderful flint work of Solutré, and the La Madeleine paintings ; after that came decline. The old hunters followed in the track of the Mammoth and the Reindeer, and reached northern latitudes, where their successors of to-day, the Eskimo, live. They left behind them the less virile types, and

FIG. 4.—Danish Midden Axe.

the early midden people lived, one thinks, in rather a kitchen atmosphere without the wit to mend their ways.

Then wise men came out of the East, and later we shall try to show how we in England were affected by these migrations. There were kings in Egypt as early as 4500 B.C., and the Mediterranean, which had seen the Crô-Magnon, and Grimaldi men, in the Old Stone Age, was to see these others who, coming from the East, or South-East, in the New Stone Age, were to press along to the cry of "Westward Ho," and build up new civilizations.

Whether the midden people died out, or were stimulated by these new-comers we cannot be sure. They had domesticated the dog, and it may have occurred to them to do the same with other animals, and so save themselves the trouble of hunting.

This we find is the next step ; man became a herdsman, and had flocks to tend. This added to his responsibilities ; while as hunter, or beach-comber, his cares were few, he must have found that with possessions his troubles began. It was necessary to find pasture for the little flock, and in the winter, no matter how hard the times were, he must keep alive some few to carry on the strain ; the animals needed guarding at night ; better pots and pans were necessary for storing milk, and in a hundred ways he was moved to bestir and adapt himself to the new conditions which arose out of becoming a man of property.

We will now turn to the geographical conditions which

confronted Neolithic man in England, and the bearing which these had on his mode of living, and the necessity that he was under of finding pasture for his flocks.

In the Old Stone Age, men walked across dry land where the Straits of Dover are now (see p. 14, Pt. I.); but as the waters rose after the last Ice Age, the isthmus across got smaller and smaller, until England was completely severed. It is probable that this did not occur until some time after the beginning of the New Stone Age, and even then the Channel would not have been so wide as it is now for a long time. This was, and still is, the great Gate into England; here have passed men of the Old and New Stone Ages, Goidels, Brythons, Belgæ and Romans, Saxons, Danes, and Normans. There have been, and are to-day, other routes, but none that can compare with the southern end of Watling Street.

We have drawn our map (Fig. 3) because we want our readers to bear in mind the physical characteristics of England; its shape; its mountains and rivers; where are the watersheds and the marshy ground. As we are going to add to this map, in each part of the series, we have drawn an England as we know it now, but readers will remember that constant alteration has brought it to its present shape. Thanet has been an island, and the Lympne Flats under water. The Wash and Fens were unreclaimed, and the East Coast by Dunwich has been steadily eaten away; there have been alterations along the South Coast and by the Isle of Wight.

In the early Neolithic days, men could stand in Gaul and look across to Kent, and say, "There is another land there like our own; there also can we walk dry foot on the hills, and find pasture for our beasts. The grass is growing brown here, let us go and see what the country is like."

On our map (Fig. 3) we have shown the chalk, and it will be noticed how closely Neolithic man kept to it. We might call them the Men of the Rolling Downs.

A drought in these early days would have led to great migrations, and the pressure from behind have forced the men on the coast to make the great adventure. The Old Testament contains the finest pictures of nomadic herdsmen. In Genesis xiii. we read how Abram and Lot returned out of Egypt, and there was strife between their herdsmen, because the land was not able to bear them, and Abram said to Lot, "Is not the whole land before thee? Separate thyself, I

Fig. 5.—Camps on Pitstone and Ivinghoe Hills, Bucks.

pray thee, from me : if thou wilt take the left hand, then I will go to the right.''

When the first Neolithic men arrived here, they would have found excellent pasture then, as now, on the Downs, and flint for their tools. They would move along the line of the old road later called the Pilgrims' Way, on the escarpment of the North Downs, secure from wolf or man. We find to-day traces of Neolithic man on this road ; there is Kitscoty to the N.W. of Maidstone ; the Coldrum monument to the W. on the other side of the Medway ; the pit-dwellings in Rose Wood near Ightham,—all dating from the New Stone Age. Neolithic man introduced sheep, goats, pigs, and cattle (*Bos longifrons*), like the small black Welsh cattle. These necessitated enclosures ; so we find along the trackways on the Downs a regular system of earthworks where men, and cattle, could be secure against attack.

It is a curious fact that nearly all the Neolithic camps are above the 500 feet contour line ; we have shaded the parts below this on our map. There was, of course, a reason for this ; not only was pasture better on the Downs, but there were fewer trees. The country was far more wooded than it is now, and man had not as yet the implements with which to make extensive clearings in the forests. It is a mistake, however, to think of these as dense tropical jungles, because the climate then was temperate, as it is now. The undrained country would have been a more formidable obstacle than the forests, and places like the Sussex Weald all sticky clay. The forests were full of wild animals ; there was the Irish elk and the wild ox (aurochs), bears and beavers, wild cats and red deer, wild boars and the wolf, and Neolithic man hunted these with dogs.

Later and more adventurous immigrants seem to have coasted round until they came to the chalk at Eastbourne. They would have set out in their dug-out canoes, as Fig. 6, and some of these have been found as long as 50 feet. On the South Downs again are earthworks and tumuli, linked up by trackways leading to Stonehenge. Others came in at the Wash, which in these days extended to where the chalk is shown on our map, and here Icknield Way goes S. to the Goring Gap on the Thames, and then by way of the Berkshire Downs again to Stonehenge. Later on Maiden Castle, near Dorchester, and its connection with the trackways, points to traffic and trade by

FIG. 6.—Dug-out Canoe.

sea. The range of Neolithic man seems to have been the Downs, the Blackdown Hills to Devon and Cornwall, the Mendips, the Cotswolds to the Northampton Heights, the South Pennines and Lincolnshire Hills, the Yorkshire Wolds and Moors, and the Glamorgan Hills, and N. and W. of Scotland, and all these parts are connected by trackways which converge on Salisbury Plain and Stonehenge, which appears to have been the richest part of England in the Neolithic and Bronze Ages, and the seat of such spiritual and civil government as there was.

It should be noted that the trackways follow the water-sheds, and so avoid the crossing of rivers—a serious obstacle to flocks and herds. In later days the great river valleys formed avenues of approach for immigrants into the country, and the fact that so many of these are on the East Coast, has rendered us peculiarly liable to invasion on that side. The tide runs up the Humber and Ouse nearly to York; up the Trent just beyond Gainsborough, and the Thames to Teddington.

We must think, then, of a gradual penetration of the country, in Neolithic times, along the various routes we have indicated, which in the end became established traffic lines because of their convenience. The first rough stockades and earth-works on the trackways developed as time went on into the hill forts we find to-day. In the later days there must have been a more ordered system of government than any tribal law which had gone before. This is forced on us by the size of the works which these people carried out, and which could

7

CAMPS

only have been possible to a people content to accept some form of control.

We must bear in mind that when we talk of the Neolithic period, we mean a state of existence which is supposed to have lasted in this country not less than 3000 years and probably longer, that is, from about 5000 or even 8000 B.C. to 2000 B.C., and it may have started considerably earlier. To realize what a very long time this was, we must remember that only 1945 years have elapsed since the birth of Christ to our own days. Neolithic man had plenty of time for the gradual beginnings which led up to the civilization of the hill forts of the trackways. Boys and girls should endeavour to see these. In our part of the world there is Icknield Way, with a contour camp on Beacon Hill, the Maiden's Bower, and Totternhoe. From Oxford you can take a 'bus to Wantage, the birthplace of Alfred, and from there climb on to the Ridgeway which runs along the Berkshire Downs. Cissbury is close to Worthing, and Maiden Castle not far from Weymouth, and every one should see Stonehenge. There is no more inspiring thing to do than walk along these trackways, which were old roads before the dawn of History as it is generally understood. If the day is hot, rest for a little while under a thorn, and then, perhaps, if you can dream dreams, and see visions, you may be able to join in spirit a party of Neolithic hunters or herdsmen journeying from fort to fort. It will be much more amusing than reading books, yet give your History a new meaning.

EUROPEAN RACES

Before we examine the works of Neolithic man in more detail, it will be as well to try and find out something about him, and the European Races during the Neolithic, Bronze, and Early Iron Ages. We can refer to ourselves as Anglo-Saxons or Britons, and yet be very wide of the mark. Assuming that we were cruising over Great Britain in an airplane, we could in a few days cover the length and breadth of the land, and if we kept our eyes open when we landed, we should find very varying types in our own country, except perhaps in the industrial areas which are pitiful conglomerations of misery.

In parts of Essex, and the South Midlands and Chilterns,

on the hills to the W. of the Severn in Worcestershire, Shropshire, and Herefordshire ; in Romney Marsh, the Weald, and the Isle of Ely, we should find a large proportion of dark-haired people with long heads, and the explanation of this is, that as these parts were off the main lines of Saxon immigration, the old British blood has lingered on. The Saxons penetrated into the country on the line of the Thames, and this element is strong in Berkshire, Oxfordshire,

FIG. 7.—Mediterranean Man.

Hampshire, Sussex, and up the Thames Valley to the Cotswolds; here you will find fair people with blue eyes. In Leicestershire and Lincolnshire are Danish types with long faces, and heads rather high behind ; high cheek-bones, and well-formed noses ; they appear to have driven the Anglians to the Derbyshire hills in olden days. In Yorkshire we find a typically English people ; shrewd, vigorous, and obstinate ; successful in business ; hard-headed and practical, yet with a great love of music. In the Shetlands, Orkneys, Hebrides, and parts of Caithness are splendid men of Norwegian descent. In the Highlands a Gaelic stock, quick-tempered and emotional ; in the Lowlands, and the eastern coast-lands, a frugal hard-working people descended from Angles, Danes, and immigrants from the E.

It is obvious, then, that our own island provides us with some very fair samples of the European races, and if we are to understand our own history, or must discover where these types have come from, this means crossing to the mainland.

The European Races have been divided into three large families or groups. The Nordic, Alpine, and Mediterranean, and the history of Europe is a recital of the migrations and minglings of these types. Nordic means Northern, and

9

FIG. 8.—Alpine or Celtic Man.

this type is sometimes called Teutonic; these people came from the steppe region to the N. of the mountains between Europe and Asia. As the climate improved after the last Ice Age this became forest. The people were tall and strong-boned, with fair hair, and blue eyes, and they were long-headed.

The Alpine people came from the mountain zone of Europe; they were thick-set, and round-headed.

The Mediterranean men came from the coast-lands of that sea; they were dark, long - headed, with oval faces and aquiline noses; of middle height, not more than 5 feet 6 inches, and the women shorter and not very robust.

The Nordic and Mediterranean types were probably descendants of the later long-headed people of the Old Stone Age, and the Alpine later arrivals from the E.

It is to the Mediterranean stock that we must look for the first of the Neolithic people in this country. It is thought that working along the coast-lands of the W. part of the Mediterranean they struck up through the Carcassone Gap between the Pyrennes and the Cevennes, at 1 (Fig. 64), and thence by way of 2, 3, through the W. of France until they came to Brittany and Normandy, then worked along the coast until they came to where the Straits of Dover now are. Remember this was not done in a day, or many days, but was a movement lasting for hundreds of years.

The later Mediterranean people were the builders of the Megalithic monuments; the menhirs, dolmens, and chambered barrow which culminated in Stonehenge, and spread from India, across to W. Europe and our own land. Megalithic is derived from two Greek words, *megas*, great, and *lithos*, stone, and its most distinctive contribution to the art of building was the evolution of the lintel ; in this detail it was allied to Egyptian and Greek building. Stonehenge is the triumph of the lintel, and the general assumption is that it dates from the end of the Neolithic, or the beginning of the Bronze Age.

FIG. 9.—The Nordic Man.

These Neolithic dolmen builders retreated before the round-headed Bronze men, who seem to have come from the Eastern Mediterranean, through Gaul to Britain. They were stalwart, dark, broad-headed men, and arrived here about 2000 B.C. It is thought that these earliest round-heads were not Goidels, and we will explain this later. It is quite possible that they may have had something to do with megalithic building, as they associated with the Neolithic long-heads ; we know this, because in the round barrows, which are of Bronze Age, round- and long-heads are found buried together. The Bronze men brought with them their flat bronze celts, as Fig. 46, and if at the first they could not manufacture these they did obtain them by trade.

About the same time the " Beaker " people arrived on the N. and E. coasts. They are called " Beaker " because of a pottery vessel found in their graves (1) Fig. 62. It may not have been a drinking-cup, but looks like one. They came from around Kiev on the Dnieper (7, Fig. 64), to the S. of the Pinsk Marshes, and then on the line 8, 9, 10, not in a month or a year, but gradually, as their numbers increased and they were forced to find new territory—in fact, just as men in recent days have gone to America to make their fortunes. These Beaker men were a mixture of Alpine and Nordic, combining the broad heads of the Alpine with the fair colouring, strength and length of bone of the Nordic. They were tall and strong-browed.

About this time we are able to find out that the conditions of life were becoming easier. The people lived longer lives, they were bigger than in Neolithic times, and there was less difference between the size of men and women.

THE COMING OF THE CELTS

At a later day, perhaps, about 700 to 500 B.C., the first of the Celts arrived; they were an Aryan-speaking people who burned their dead. Here we might explain what is meant by the Aryan-speaking peoples, because the spread of this language is one of the wonderful things in the world's history, like the La Madeleine painting. The Aryan language is also described as being Indo-European, Indo-Iranian, and Indo-Germanic. Towards the end of the eighteenth century, similarities were noticed in the construction of languages seemingly so different as Sanscrit, Greek, Latin, German, and Celtic, and later all the European languages, except Turkish, Finnic, and one or two others, were added, with some modern Indian languages, to a group which has been derived from this primitive Aryan tongue. This does not mean that all the millions of Aryan-speaking people to-day are descended from Aryan stock; what it does point to is some wonderful idea which spread across Europe like a flame burning dry grass.

The exact spot where the original Aryans lived is still a matter of debate : one idea is that it was in South Russia or Hungary; another, on the Iranian plateau to the S.E. of the Caspian Sea. From there the language spread S E. across the Indus into India. The route to Europe may have been to the E. of the Caspian Sea and then W. across the Volga, Don, and Dnieper, to 7 (Fig. 64), whence came the Beaker people. Or N.W. from the Iranian plateau, and S. of the Black Sea into Asia Minor and the Ægean. Now

FIG. 10.—Flint Miners.

language does not spread as a fashion, but because it is the vehicle of thought embodying a dominating idea.

The diffusion of the Aryan language coincided with great changes and migrations of the European peoples. The old Neolithic civilization had carried men forward as a tribe, and in a state which did not offer much opportunity to the individual. While the pioneer work was being done, the adventurous men had plenty to occupy them, and then may have become restless as conditions became more settled, and have seized power, not necessarily from a selfish point of view,

FIG. 11.—Deer-horn Implement.

but to satisfy wider ambitions and to obtain more movement and colour in life. We come to the Age of Heroes. The chieftain, or patriarch of the tribe, has to give way to the hero, who welds it into a nation and becomes a king.

The Celts, an Aryan-speaking, fair-haired people began to come over from the Continent about 500 B.C. bringing with them the first weapons and implements of iron. They spoke two kindred but slightly differing tongues which still persist in these islands in forms which, in the main, are not greatly altered. They were called, according to this division of speech, the Goidels and Brythons, and by Roman writers the Gauls and Britons. The descendants of the former are the Irish, the Highlanders of Scotland, and the Manx, of the latter the Welsh and the Cornish.

About 75 B.C. came the Belgæ, of Celtic stock with an admixture of Germanic, and Cæsar found them in the possession of the S.E. districts.

FLINT IMPLEMENTS

Having now given an outline sketch of the various peoples we shall meet with in this book, we will go back to the first of these, the men of the New Stone Age. We will examine first their implements, and then later consider the work they did with these tools. These Neolithic implements are not necessarily of polished stone, as some people seem to think. Flint was still chipped as in the Old Stone Age: sometimes it was chipped and ground, or polished in parts; sometimes completely so. We can only give a few of the more typical

FIG. 12.—Deer-horn Implement.

14

implements, and we strongly recommend our readers to pay a visit to the Prehistoric Room of the British Museum, where the endless variety of the implements can be studied in detail. Neolithic implements are found on the surface of the ground or just under it, and are not dug out of gravel as those of the Old Stone Age.

When our readers pass on to the standard textbooks of archæology, they will be meeting constantly such terms as nucleus or core, flake, and bulb of percussion. It may be as well to explain these. Flint is dug out of the chalk in

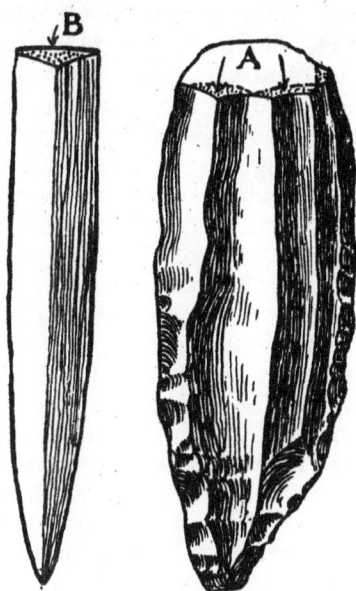

FIG. 13.—Flint Flake and Core.

separate blocks or nodules complete in themselves ; not cut out of a mass, as in the case of stone and rock. At Cissbury near Worthing, and Grimes' Graves near Brandon in Norfolk, the pits formed by the early miners to obtain their flints have been discovered, and it is thought the implements were roughly finished here for export. They used deer-horn picks, and shoulder-blades as shovels, as Fig. 10. These can be seen in the Prehistoric Room at the British Museum, with horn punches and chisels, as Figs. 11 and 12. The flints have a white skin called the crust, and the old men often left part of this on the implement. Remember they had not any metal hammers, and that a rounded pebble was used instead. The first step was to knock off the top of the nodule, so as to provide a flat table at A, Fig. 13. This tabular surface was held nearly at a right angle, and the flaker with his pebble struck a sharp blow a little back from the e!ge at the arrows, on the line of the intended fracture. By long practice he knew exactly the position and force of the blow necessary to detach the flake; it is obvious that he might obtain one of triangular section from the left-hand arrow as at B; this

15

would have a mid-rib up its centre, and two keen cutting edges, and be useful as a knife or lance-head. From the right-hand arrow he would obtain a flake with two ribs up the middle; it was this type of flake, cut up into short lengths, which was used until recent days for flint-lock guns, and strike-a-lights. It is becoming increasingly difficult, in these mechanical days, to appreciate the manual dexterity of the old workers, who were content to regard the hand as the most wonderful tool of all. Try and make a flint implement yourself, but wear motor goggles to safeguard your eyes, and you will leave off with a new respect for these old handicraftsmen.

The block from which the flakes are struck off is the nucleus or core, and in the Prehistoric Room in Table Case A, you can see one with all the flakes replaced. In the Gallery over are cores from France called, by the peasants who find them, *livres de beurre*, or pounds of butter.

Flint is a curious material, intensely hard, it is yet rather elastic. When it is struck by the hammer-stone, the blow detaches the flake with part of a cone under the point of impact; this is the bulb of percussion, and is generally regarded as a sign of human work on a flint. The implements resolve themselves into two types. First these made from the core itself, the flakes being removed to give the desired shape. Naturally the larger implements, like the hand-axes in Part I., and the celts, axes, and hammers, in this part are shaped cores. In the other type flakes were struck off the core and were used for knives, lance and arrow heads,

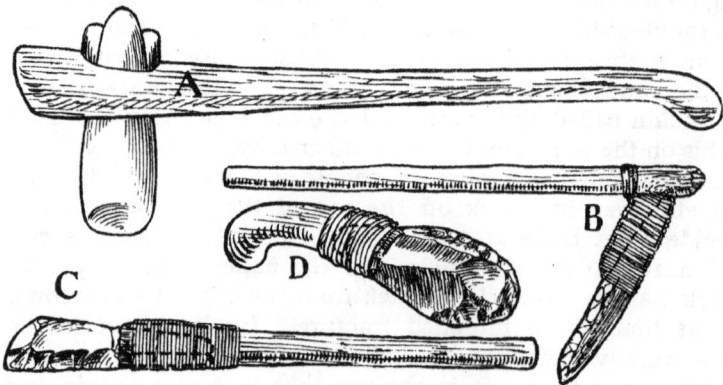

FIG. 14.—Hafting of Flint Implements.

16

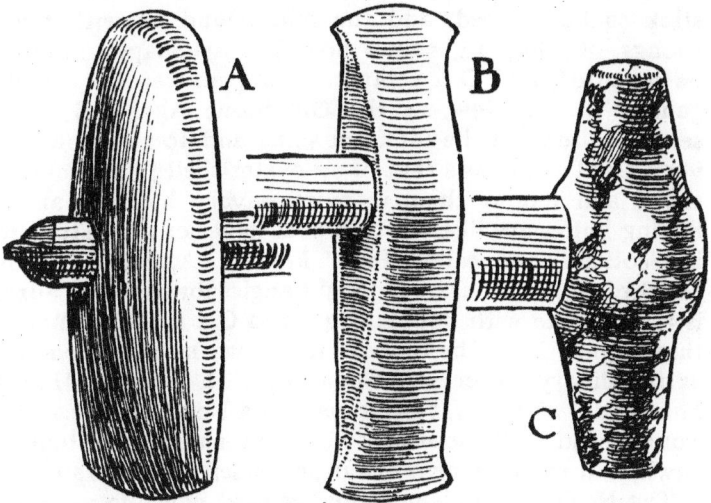

FIG. 15.—Stone Axes and Hammers.

scrapers, borers, and all the little odd tools which were so useful.

Fig. 14 shows a few typical implements, and the way they were hafted or had handles fitted. A is the celt, or axe, and is the Neolithic descendant of the hand-axe of the Old Stone Age. Celts have been found varying from an inch or so long up to 15 inches or 16 inches, and were the most important implements of Neolithic man. They were driven into the head of a wooden handle as at A, and then wedged from the top. Sometimes the celt was fixed into a deer-horn socket driven into the wood. With celts trees were cut down and all the rough carpentry done. The stone celt or axe was the forerunner of the bronze celt, and led to the iron axe which has been one of the most useful tools to man throughout the ages. A, Fig. 14, shows a polished stone celt. These at first were chipped out of flint. Then the cutting edge was ground, and finally the whole celt polished. B, Fig. 14, shows a rougher, unpolished type, hafted at right angles to the handle for use as an adze; this may have been used like a hoe to chop towards the foot, and must have been very useful in making dug-out canoes. Rougher stones mounted in this way were used perhaps as hoes for agriculture. Early man cultivated the terraces or lynchets near their encampments, as Fig. 2. For this method of hafting any branched stick could be used, and the flint bound on with raw-hide thongs. C, Fig. 14, shows how a chisel-shaped flake could be mounted, and D a scraper. Scrapers were as useful and general in the New, as the Old Stone Age, and probably served to remove the fat from skins and to scrape wood. A very usual shape was that of an oyster-shell ; the Eskimo use these, and mount them in morse—ivory handles, and their flaying knifes are like the thin oval flakes of greenstone found in Scotland, and called Picts' knives. A, Fig. 15, shows a polished stone celt hafted at right angles for use as an adze. B is a stone axe with double edge, and C a stone hammer. In thinking of how these were made we must remember the extraordinary patience of the savage. Lafitau, in *Mœurs des Sauvages Américains*, 1724, says that a North American Indian would spend all the leisure of his life in making one stone toma-hawk, and we may, or may not, consider that a waste of time

The Neolithic implement maker used volcanic rocks for his axes, and after roughly trimming these to shape, finished by grinding the axe on a grindstone, not one that turns round, but by rubbing the axe on a stone, as the carpenter sharpens his plane iron. The boring of the hole was done last, with a stick, or hollow bone, and sand and water. Any sand hard enough to scratch the stone would cut the hole in time. The drill could have been turned with a bow, as Fig. 47, Part I. Odysseus drills out the eye of Polyphemus by means of a stake with a leather thong around it, "as when a man bores a ship's timber " (*Homeric Greece*, p. 64).*

Some of the stone axes have one edge and a rounded head, and may have been used for splitting wood, by hammering the head with a wooden mallet. Others have a purposely blunted edge, as if for use as battle-axes, with less chance of cutting the wielder, and just as much power to damage the enemy. Amusing traditions have gathered around the old stone celts ; the country people in the past thought they were thunderbolts. Stone hammers were known in Scotland, until the end of the eighteenth century, as Purgatory Hammers, and were supposed to have been buried with the dead, so that they could hammer on the gates of Purgatory, till the heavenly janitor appeared. Another point to be remembered, and one which we have so often emphasized, is

* Later references to *Everyday Things in Homeric Greece* by the same authors, will be given as *H. G.*

that stone continued to be used after the advent of bronze. Sir William Wilde, writing in the *Catalogue of Stone Antiquities in the Royal Irish Academy Museum*, stated, in the middle of the nineteenth century, that stone hammers and anvils were used by Irish smiths and tinkers, until about that time. Again, Sir John Evans, in *Ancient Stone Implements*, published in 1872, says that up till that time flints were sold in country shops for use with steel to make fire. Leaving the larger implements, we can turn to the lance, javelin, and arrow heads, and the many things which were made out of the flakes. We have seen by Fig. 13 how the flaker went to work. Long flakes up to 8 and 9 inches were possible,

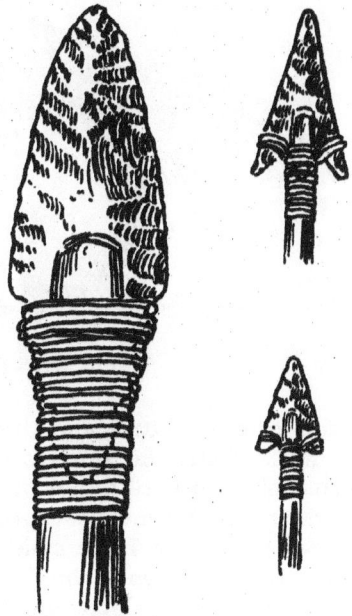

FIG. 16.—Flint Spear and Arrow Heads.

and these were used for lance-heads ; shorter ones for javelins and arrows ; thicker and rougher flakes for scrapers. Having obtained the flakes, the maker then proceeded to trim these into the desired shape, by what the archæologists call secondary flaking. Some of this, as in the Danish specimen, in Case 134 in the gallery of the Prehistoric Room at the British Museum, is rippled along the edge of the implement in a most delightful way. Opinions are divided as to how this secondary flaking was done. A flint punch, or fabricator, may have been used ; or the flake held flat, face uppermost on an anvil stone, may have been trimmed by hammering tiny flakes off the edge with a hammer-stone. The Eskimo place the flake over a slight hollow in a log, and then press an ivory tool which spalls off small flakes. Capt. John Smith, writing in 1606 of the Indians of Virginia, said, " His arrow-head he maketh quickly with a little bone, which he ever weareth at his bracert (guard on wrist against bow-string), of any splint of stone or glasse

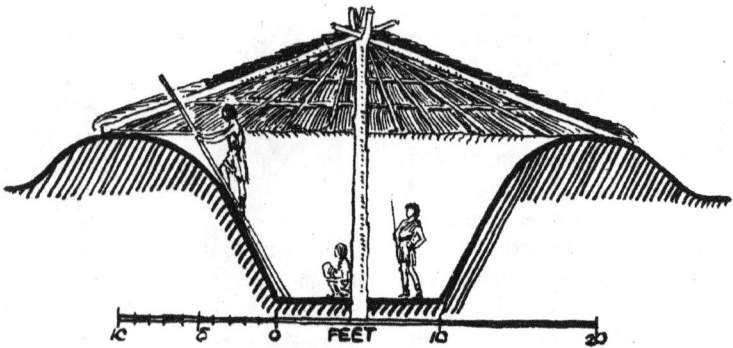

FIG. 17.—Pit Dwelling.

in the form of a heart, and these they glew to the end of their arrowes. With the sinewes of deer and the tops of deer's horns boiled to a jelly, they make a glew which will not dissolve in water." This means a form of mounting as Fig. 16. The arrow-heads must have called for wonderful handling when being made. As with the Celts, tradition has gathered round the arrow-heads, which, until quite recent times, were called elf-darts by the country people, who thought that the fairies used them to injure cattle.

HOUSES

Having seen something of the tools which Neolithic man possessed, we can pass on to the work he did with these, and will begin with the houses he built. In Fig. 5 very simple huts are shown which resemble those of the Old Stone Age shown in Fig. 59, Part I. It is a type which has always been used by primitive man, and we can remember charcoal burners in Kent who housed themselves in this way. This would be the hut, of what is called the hut circle, that is, the shallow depressions which are found in Hayes Common in Kent, and many other parts of the country. The hole which remains now is dished out like a saucer, because in the time which has passed the outer edges have been trodden and washed down by the rain. Originally the hole was dug out and the ground heaped up around ; this would have given headroom inside, and have

taken the place of the
vertical walls that
came later on. A
central roof-tree sup-
ported the saplings at
the top, which, resting
on the bank at the
foot, formed the roof.
A rough thatch com-
pleted the whole. Very
much deeper pit dwell-
ings were formed, as
Fig. 17, in the same
way, and these suggest

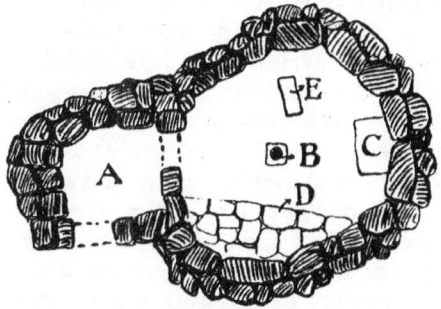

FIG. 18.—Plan of Hut.

that fear prompted the form of their construction. It is
obvious that this type was not very noticeable to prowling
enemy bands, and the wolf would hesitate to leap down into
such a trap. The pit dwellings are thought to be earlier than
the shallower huts, and were only possible in a dry soil;
this obtained, they were doubtless warm in winter and cool
in summer. The cooking hearths, as on Hayes Common,
often took the form of small pits outside the huts. A fire
was made in these with large stones in it, and the ashes being
raked to one side, the carcase was placed in the pit and

FIG. 19.—Neolithic Hut.

covered over, when the heat of the stones turned the pit into an oven and cooked the meat. It is very probable that the accidental introduction of ore with the fuel into one of these hearths led the way to metal smelting. The floors of the huts would have been covered with bracken, like straw in a stable, and carpet-sweepers were not needed.

Fig. 18 shows the plan, and Fig. 19 the outside of an interesting development from Grimspound, Hambledon, Dartmoor. Here are the remains of twenty-four huts, surrounded by a double wall enclosing about 4 acres ; quite a little village. The roofing of the huts was on the same principle as Fig. 17, but of course all this has long since gone. The plan is interesting because the hut has now developed a porch or outer parlour at A, which must have added to the comfort of the inhabitants ; at night it may have been used as a stable. The house is rising up out of the ground, and has rough vertical walls ; at the entrance the builders selected upright stones for the door jambs, which are covered with a stone lintel ; this is an important detail and links the house up with Stonehenge, as we shall see later. The hut is about 11 feet diameter inside, with an inside hearth for the fire at C, and a cooking-hole at E ; there is a raised dais at D paved with flat stones, about 8 inches higher than the general floor. Here the family could sit on bracken and fur rugs in great comfort. The central roof-tree, supported on a stone at B, would have been used, like the pole in an army bell-tent, to hang things on. As late as Cæsar's time the Gauls squatted in straw around a low table, and tore their food like animals, using their fingers and only occasionally their knives.

Flint thumb-scrapers found in the Dartmoor huts suggest skin clothing ; though weaving appears to have been started in the Swiss lake

FIG. 20.—Strike-a-light.

FIG. 21.—Flint Sickle.

dwellings in Neolithic times, it is doubtful if it started here till the Bronze Age. Very few ornaments have been found in long barrows.

Skin clothing does not necessarily mean that Neolithic men only wore the rough pelts of animals; we saw in Part I. how the women of the Old Stone Age could make very good bone needles, and a visit to the Ethnographical Gallery, at the British Museum, will show us what beautiful skin garments the Eskimo can make. Neolithic garments may not have been quite as well made as these, and in Fig. 56 we have shown the man and woman of this period, on the left of the drawing, in a simpler type of clothing. The Picts, who were descendants of the Neolithic men, tattooed themselves, so this method of decoration may have gone back to the New Stone Age.

FIG. 22.—Grinding Corn.

Fig. 20 shows a way that the Neolithic woman made fire ; a piece of flint was used, in conjunction with a lump of iron pyrites, as a strike-a-light. Pyrites is found in the lower chalk beds, and may first have been used as a hammer-stone on flint, when the resulting sparks would suggest its use as Fig. 20. The sparks falling on dry moss could be blown into flame. Very beautiful flint knives, as Fig. 21, have been found, and it is thought that these were used as sickles. The reaper gathered the ears of the corn in one hand, and cut these off as shown. We have already referred to the lynchets found on the Downs which are supposed to be cultivation terraces. When the corn was cut the threshing was a very simple business, and then came the grinding into flour. Fig. 22 shows a saddle-back quern : the grain was placed on this, in the hollow made by use, and the upper stone pushed to and fro until the corn became flour. Neolithic man would hardly have been able to obtain yeast, and probably his bread was unleavened, or the flour mixed with honey and baked into biscuits. Fig. 23 shows a pot quern, like a modern pestle and mortar, which would have been very useful for pounding things up. These querns were made of gritstone, and can be seen at the British Museum in Wall Case 5 in the Prehistoric Room.

POTTERY.

We come now to one of the most important discoveries of Neolithic man or woman ; he or she found out the way to make pottery. Fig. 24 shows a bowl of thick dark ware made without the potter's wheel, probably in the same way that the Akikúyu of British East Africa work to-day. These people temper their clay by pulling it into small pieces and freeing it from stones ; it is then dried in the sun, and after mixed with water until it is plastic. A fine sand is then kneaded into it, in the proportion of about half in half, and the clay finished in long rolls. One or two of these are formed into a collar shape, and with one hand inside this, and the other out, it is gradually modelled into the shape of the top half of the pot, more clay being added in rolls as the work proceeds. The half pot is allowed to dry in the sun for some hours, except the lower edge where the join has to come ; this is protected by leaves. This edge has rested on leaves while the top half was being made so that it could be turned more easily, and this movement must have suggested the potter's wheel later on. In the next stage this top half is turned upside down on its already finished mouth, on more leaves, and

FIG. 23.—Pounding Grain.

ABOUT 6¾"

FIG. 24.—Neolithic Pot.

FIG. 25.—Making Pottery.

the modelling proceeds as before, more material being added as required to form the bottom, the shape being given by one hand in, and the other out, until there is only room for one finger, and then the hole is closed, and the pot finished. Again, a few hours are allowed for hardening, then the pots are placed mouth downwards on the ground, and a bonfire of brushwood made all around them ; when this has burned out, and the pots are cool, they are ready for use. The only tool used, beside the hand, is a piece of gourd shell.

Fig. 25 shows how Neolithic woman went to work, and Fig. 26 a pottery spoon she made, which can be seen at the British Museum.

WOMAN AS AN INVENTOR

The Akikúyu pottery is made by women, and the probability is that Neolithic woman did this work, and looked after the home, while her husband was hunter and herdsman. She probably did far more than just cook and mend ; we must think of her as an inventor. With pottery the long train was started which has led up to the modern saucepan ; before then, meat could only be roasted over a fire, or baked in a cooking-pit, but with a stout earthen pot that could be placed in the ashes the Neolithic equivalent of Irish stew was possible Water could be heated, and milk and grain stored.

It will be noticed that the pot shown in Fig. 24 has a rounded bottom, which suggests that it was blocked

up on two or three
stones, and a fire made
under it.

Perhaps it was the
woman who noticed that
cattle ate the seeds of
grasses, and experimented
by grinding some between

FIG. 26.—Pottery Spoon.

stones; she may have tasted the flour and found it
sweet, and then have brought home more seeds. A few
seeds blew away into the ground newly turned up at the base
of a hut, and the woman watched these growing and watered
and tended them. In this way it may have occurred to her to
make a garden, and she discovered that cultivation improved
the crop; once this fact was appreciated there were endless
opportunities; the crab apple, wild plum, and other fruits
could be experimented with, and most probably woman was a
gardener before man became a farmer; of one thing we may
be quite sure, Neolithic man did not rise up one day and plant
an acre lynchet, without endless experiments and questionings
going before.

If Neolithic woman made pottery, then it is to her we
must give the credit for a renaissance of the Arts. There
had been a great slump in the art world since the La Madeleine
times of the Old Stone Age, but with the coming of pottery,
pattern began. At first it did not amount to much more
than cutting lines in the damp clay, or denting it with the
finger nail; still it was a start, and before this book ends
we shall see how in late Celtic times pattern became very
beautiful.

NEOLITHIC EARTHWORKS

Having seen something of men's houses in Neolithic times,
and the more domestic details of their lives, we can turn to
their larger works. The trackways, or road system shown
on Fig. 3, link up a series of splendid earthworks, and many
of these are of Neolithic construction. Starting perhaps as
simple cattle enclosures, surrounded by a ditch and bank,
with some additional precautions taken at the entrances,
these camps were gradually improved, until we arrive at such

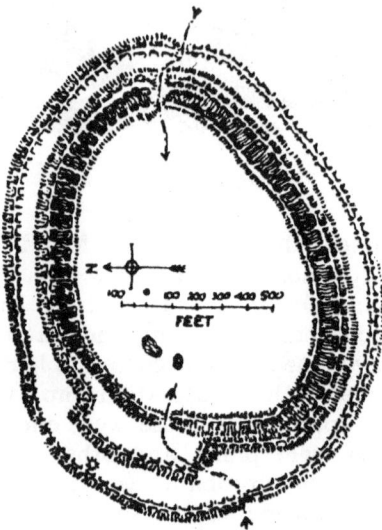

FIG. 27.—Plan of Badbury Rings, Wimborne, Dorset.

FIG. 27A.—The Eastern Gate at Badbury Rings.

FIG. 27B.—The Banks at Badbury Rings.

a masterpiece as Maiden Castle near Dorchester, More banks were added. the entrances made into mazes of ingenuity, and the whole developed just in the same way as the Tower of London, where we find the Norman keep surrounded by much later works.

It is very difficult to estimate the age of earthworks ; especially the very simple ones. Neolithic flint implements and pottery have dated some ; in others Roman coins have been found, but this would not justify us in saying that an earthwork was Roman. The Romans fortified their camps when on the march, but did not of course ever live in hill forts. Roman coins in these may point to the times of the Saxon terror, when the Britons fled to these forts as places of refuge and took their money with them.

Earthworks are classified by archæologists as A, Promontory Fortresses, where a piece of high ground inaccessible by reason of precipices or water on one side, has been defended by artificial works on the other. B 1 are Hilltop Forts with artificial defences following the natural lines of the hill, and are sometimes called Contour forts. B 2 are Forts on high ground, less dependent on natural slopes for protection, and there are later types which do not concern us now.

To illustrate the general principles of this method of

fortification by earthwork we have chosen Badbury Rings, near Wimborne, Dorset, which is classified under B1, and the plan of this is shown on Fig. 27. It may be well to give first a brief description of the terms used in describing an earthwork. Vallum, Rampart, and Agger, all mean earthen walls, see 1 on section on Fig. 27A. Fosse or Ditch at 2, Escarpment is the slope at 3. Counterscarp at 4 ; if the counterscarp is brought up above the level line as a smaller rampart, this is a revetment. The flat piece of undisturbed ground at 5 is a Berm. The plans of earthworks, which generally look like hairy caterpillars biting their tails, show the top of a slope as a thick line tapering off down the slope.

Now as to the way the old builders went to work. To start with, they had as good an eye for the possibilities of a piece of country as a Royal Engineer officer, or a fox-hunting squire. They always chose pleasant sunny situations where the thyme-scented grass gave good feeding for their cattle, and the scabious flowers nodded in the breeze to the song of the skylark. There is no more pleasant place in which to loaf than an old earthwork ; you can always get into the sun and out of the wind, and the slope of the banks is exactly right for an easy position from which to gaze over the country-side, and that is just what the old men wanted to do. Their cattle would have grazed on the hillside, meanwhile the watch-man kept a look out for wolves and wild boar, or wandering cattle-lifters. Cattle was wealth in those days.

The builders then chose the rounded hump of a chalk down, which was not controlled by any higher ground, and it is probable that the first thing they did was to dig one simple ditch and bank, or fosse and vallum. In doing this they had to use antler picks and shoulder-blade shovels, as Fig. 10 ; remember they had no metal as yet. They doubtless carried up the chalk in rough baskets, and so raised the bank above them. On examining an old earthwork, the first thing to do is to discover the natural level, as dotted line on section on Fig. 27A, and then see how they went to work, because at first sight the fosses are so deep, and the banks so high, that it seems impossible such work could have been done without steam navvies. When we have found the natural level we discover that the art of the job was, that by the basket of earth dug out, not only was the ditch lowered, but the bank raised ; see A, Fig. 27B, and that a higher bank was made more speedily on a slope as at B, than on the level.

GATEWAYS

Again, on a very steep slope as C, the soil dug out could be thrown downhill.

Still, notwithstanding all this, these earthworks must have been tremendous undertakings. The outermost of the three banks at Badbury, which we illustrate, is 1 mile in circuit; at Maiden Castle, near Dorchester, nearly 1½. Particular care was given to the design of the entrances. At Badbury there are two, one on the E. and the other the W.; the dotted line shows the way in. On the W. side the banks have been cut through in other places in recent times, but in old days any invading force had to come as the dotted line, which left them very much at the mercy of the bowmen on the banks above them. A " flanking " entrance was so arranged that the right side (unprotected by shield) was exposed to the defenders' arrows. The tops of the banks were palisaded, and the bottoms of the ditches were perhaps filled with sharpened stakes. The wide areas between the banks, called " berms," may have been used as cattle pens ;—a stampede of half-wild cattle at night would not have been pleasant ; or, as at Maiden Castle, the camp may have been divided into two parts for the same purpose.

Hut circles are found in the earthworks, which suggest huts as shown in our drawings. Heaps of sling stones have been found, and bracers, or wrist-guards, which show that bows were used.*

Figs. 28, 28a and 29 show the remarkable and extensive early Bronze Age tumulus of Bryn Celli Dhu, mentioned in the Introduction. There was a large outer circle of standing stones (now gone), This was in full view. The three other circles were hidden in the mound covering the tumulus. The two next circles indicated by heavy black and light broken lines (Fig. 28) consist of upright stones with dry stone walling between them. Actually they are not circles for, when traced out from end to end (including the passage and chamber in the heart of the mound) they are found to form a continuous spiral—a magic cypher! The innermost circle is of single stones. At the foot of some of them the burned bones of young persons were found. The dotted lines connecting them show how they are arranged in opposite pairs in line with the centre of the whole tomb. To one side of the inner chamber will be noticed a recess to make room for the black

* A further note on the hill-forts based on recent discoveries will be found on pp. viii & ix.

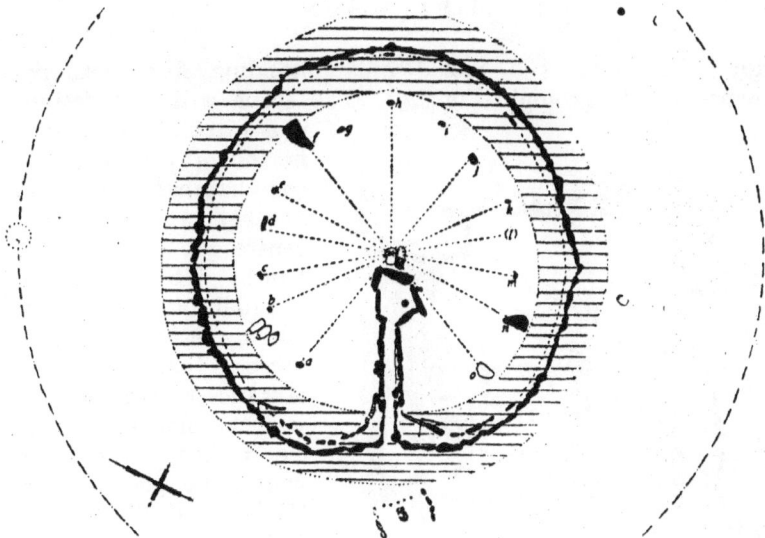

FIG. 28.—Plan of the Chambered Cairn of Bryn Celli Dhu, Anglesey.

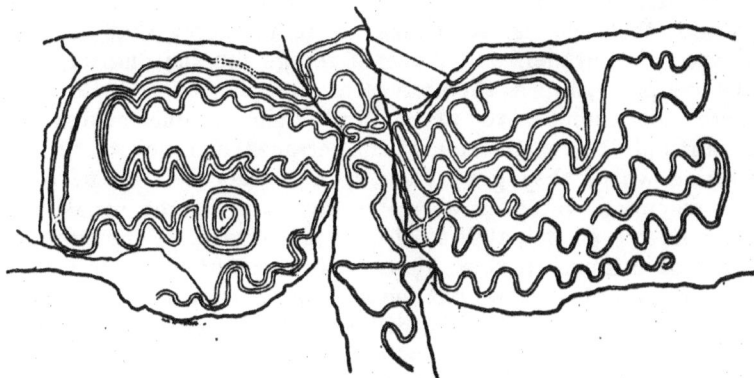

FIG. 28A.—Diagram of Markings on the Pattern Stone, Bryn Celli Dhu.

dot. That dot is a tooled pillar standing five foot six inches above ground (Fig. 29). The stone with the design illustrated in Fig. 28a is shown just beyond the chamber and by it another at the very centre of the whole tomb. This covered a small pit containing two large lumps of red jasper. Just outside the entrance to the passage was the skeleton of an ox, its head turned towards the portal. The Ministry of Works has restored the chamber and the covering-mound. The date is about 1500 B.C., and the monument seems to represent the amalgamation of many cults of the circle and the barrow.

FIG. 29.—Pillar Stone in Chamber of Bryn Celli Dhu

WATER SUPPLY

There has been considerable discussion as to how the Hill Fort men provided themselves with water, and there are various theories. First, it must be remembered that the fort formed the citadel, and place of refuge for the district, and the people grouped themselves around it. Their little huts were not difficult to make, and their simple husbandry meant only the cultivation of the terraces, or lynchets, on the hillside where they grew their corn ; they did not need or use so much water as we do to-day, and in the usual way were free to go downhill to the nearest stream.† The country was not drained in those days, which meant water lay on a higher level than now, but leaving springs on one side, there is the dew pond which is still used to water cattle on the Wiltshire Downs. This is made as Fig. 30. A shallow saucer-like depression is cut in the chalk, and lined with straw. On

† At Maiden Castle excavation revealed an intricate system of runnels and gutters on the chalk floor converging on certain pits. These pits may have been lined with sewn skins to make watertight cisterns and the gutters possibly puddled with clay.

FIG. 30.—A Dew Pond.

this comes a layer of puddled clay, with rims of chalk to protect the clay from the feet of cattle. Loose flints are put on the bottom, and the pond is started with a little water in it. The straw and clay cut off the heat of the earth, and when the moist mists drive over the Downs at night and come to the cooler pond, they condense on its surface. Ordinary ponds are formed in this way, where a pocket of clay comes in a warmer soil. Water drains into it, and the cattle puddle up the clay till it is free from cracks and watertight, and so the pond extends.

In the hot summer of 1921 we were going through Dorset looking at earthworks, and found the pond on the top of Holt Heath, near Bull Barrow, full of water, while the Tarrant river in the valley close by was absolutely dry. The Wycombe chairmakers, who go into the woods to turn chair legs, obtain water in an ingenious way. If you examine the bole of a beech tree, you will find well-marked channels, where the rain and condensed dew runs down the tree-trunk. The chairmaker makes a cross cut in such a channel, and drives a chip of wood in which diverts the water into a pail ; turning on a tap is not the only way to get water.

LIFE IN THE WILD

We think other questions may have occurred to boys and girls who have visited a hill fort ; they may have asked themselves, how early man withstood the cold and rain in such an exposed position, with only very scanty clothing. The Great War was a revelation as to the amount of hardship modern man could withstand, and yet remain healthy, but a happier example was·given by a Mr. Knowles in 1913. Mr. Knowles is an American ; born in the backwoods, he ran away to sea as a boy ; later he was a trapper and guide, and now is an artist. Without knowing anything about primitive man, Mr. Knowles wondered whether it would be possible for a modern man to go into the wilds and

LIFE IN THE WILD

support life without any outside aid ; to depend entirely on one's own effort. He determined to try, and on 4th August 1913 walked out alone into the woods of Northern Maine, naked, without any weapons, tools, knives, or matches. His book *Alone in the Wilderness* tells us how he fared. Fire was made with a fire-drill, as Fig. 47, Part I. ; and the inner bark of cedar braided into thin rope used for the bowstring, until later, when game had been killed, sinews were available. A log too heavy to move, was cut into short lengths by lighting fires at the places where it was to be divided ; sticks were pointed by burning the ends and then scraping away the char. A maple had fallen on to a hornbeam and smashed it up, and this provided the slivers of wood which could be scraped down with a " sharp rock " into the bow and arrows. Food was toasted over a fire, or on rocks heated by fire, and the fire banked down lasted for days. Mr. Knowles found it quite possible to walk about naked by day, but needed leg coverings as a protection against briars, and a rug for the night ; in this he was like the Australians and Tasmanians (p. 35, Part I.). The rug was obtained by trapping a bear in a combined pit and deadfall trap. Pointed stones and digging sticks, as Fig. 62, Part I., were used to dig the pit, and the bear when caught, killed by a blow on the nose from a hornbeam club. We may be quite sure that prehistoric man used all sorts of traps and snares in this way. Mr. Knowles used sharp stones for the skinning, and " quantities of meat came off with the skin " ; this gives us a clue as to why prehistoric man used so many scrapers. Some of the bears' meat was smoked for keeping, and all the sinews kept for ties. There were blueberries and raspberries for the picking ; various buds and barks were chewed, and frogs eaten, but not liked. Trout were caught by breaking down a beaver dam, which lowered the stream above, and left the fish stranded in pools. Animals were surprised in the act of killing, and driven off their prey ; an otter who had killed a trout ; a bear, a deer. Mr. Knowles did not suffer from the lack of salt, except that his food was not so palatable. For huts rough shelters were made, like Fig. 37, Part I., and mocassins were made of the inner lining of cedar bark, until skins could be obtained. Bowls, in which water could be heated, were made of birch bark skewered into shape, and these do not burn below the water-line. Mr. Knowles' book is illustrated by drawings made with char-

FIG. 31.—A Deadfall Trap.

coal from his fires on birch bark; he actually contemplated painting, and started making paper and brushes.

He passed his forty-fourth birthday in the woods, and was examined by Dr. Dudley A. Sargent, the physical director of Harvard University, both before and after his experiment. According to the system employed at Harvard, his physical condition equalled 876 points before, and 954 after. If a twentieth-century man could do all this, we do not think there is any need to be sorry for prehistoric man in his hilltop fort; the sun and rain would not have worried him, and he probably thought of himself as being tremendously up to date. Mr. Knowles feared the cold, but found that the real trial was the isolation from his fellow-men. This seems to us a very just conclusion, and has been proved over and over again. Where an individual, or race, is cut off, then development is arrested; however, in this book we are concerned with communities which are continually increasing in size.

SOCIAL LIFE

The concentration of a number of people either making or living in a hill fort was to have great results. In the old days, the hunting tribe was like a large family, who very speedily knew all one another's good points, and were so apt to emphasize the bad ones; life was not at all exciting. The keeping of

CIVIL

cattle brought more people together, and the simple enclosures developed into places like Maiden Castle. Here there must have been a bustling life, with all kinds of men coming and going, and new things to be discovered. Think of the excitement caused by a trader from overseas, arriving at Weymouth, and trudging over the hills to Maiden Castle, and bringing the first bronze celt; the hubbub that would have arisen among a people who had never seen metal before. Customs would arise, and Law solidify out of these. Language would develop around the hut fires, and traditional tales form the beginnings of literature. These hill forts are evidences of a more ordered system of life than anything which had gone before; even to-day with all our transport system, and organized labour, the construction of either Badbury or Maiden Castle would call for concentrated effort. To make a flint implement, which you do yourself, is one thing; to construct a camp which needs the labour of many men is quite another. It had to be planned; there must have been some few men who were skilled in the design of camps, and could say to the tribesmen, "To-day we will cut this ditch, and dump the stuff here to form a bank. You are going wrong there; and you have not allowed sufficient room for that escarpment, because the angle of repose at which chalk will come to rest is flatter than that," and so on.

Whether they were made by slave, or free, cannot now be ascertained, but probably by freemen. The beginnings of slavery are to be found in war, and it is doubtful if the tribesmen were sufficiently organized as yet to combine for warfare; the forts would have had to withstand raids, not endure sieges. Combination for the arts of peace leads in the end to the application of the same principles to war; prehistoric man probably first massacred his captives, until it occurred to him as being wasteful. when they would have been enslaved instead.

If our readers will read Mr. Hippisley Cox's book, *The Green Roads of England*, they will find how these hill forts are all linked up on a trackway system, as well adapted to the needs of the time as the Roman roads and stations later on. This road question brings up fortification, and what it means. Let us imagine Badbury, not grass grown as it is to-day where with a tea-tray we can toboggan, but all shining white where the chalk banks had been thrown up; or Maiden Castle, $1\frac{1}{2}$ miles round its outer circuit. It must have been startlingly formidable in appearance. As the later tribes came in as immigrants, and found their way along the trackways, these hill forts were there to bar their way. Of course, there were not any invading armies in those days, who needed to maintain lines of communication with the coast; the invaders were tribes who wished to settle down. In the case of hostile tribes, they certainly could not afford to cross a trackway and leave a hill fort on their flank or rear, unless they came to terms with its inhabitants. In this way these hill forts played exactly the same part as the Norman Castles and walled towns of the Middle Ages.

LONG BARROWS

We can now pass on to the Neolithic Long Barrow, or Burial mound, because, apart from its spiritual significance which we will discuss later, it has great interest in its structure. The Long Barrow derives its name from the fact that it is egg-shaped on plan, and there are two types; those having chambers inside for the interment, and others where the bodies were covered directly by the earth; these latter have a ditch at the sides leaving a wide path at the original level at each end. Generally placed E. and W., the burial is usually in the E. end, which is higher and broader than the W. It is a curious fact that the Neolithic long-head built a long

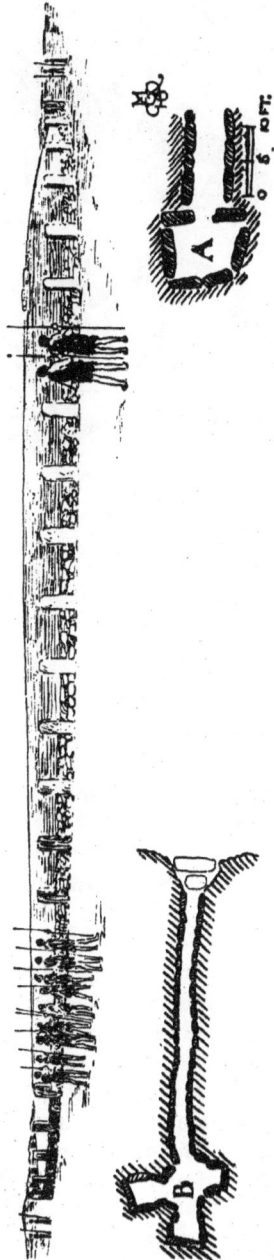

FIG. 32.—Neolithic Long Barrow, restored. At West Kennet, near Avebury, Wilts.

37

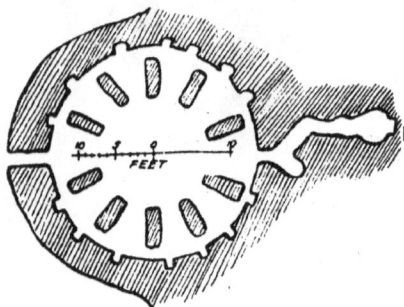

FIG. 33.—Earth House, Usinish, South Uist, Hebrides.

barrow, while that of the later round - headed Bronze man was round.

Fig. 32 shows the West Kennet long barrow. Originally it was about 336 feet long by 75 feet wide at the E. end, where it was some 8 feet high. The small figures at each end are in scale with the length, and serve to give an indication of its size. The sepulchral chamber, as the plan at A, was about 60 feet from the E. end, with an entrance corridor from the outside. It is the construction of this chamber and corridor, with large stones, which makes it a megalithic structure, and so links it up with Stonehenge. The building principle is the same, large stones are placed on edge, and the covering formed by others laid flat as lintels. In other structures of this sort, where the span was too great for one stone, courses of masonry were projected from either side as corbels, until the central space was narrow enough to be bridged. See Picts Houses, Figs. 33 and 34. This is the same method of building as that employed in the Tomb of Agamemnon (see *H. G.*, p. 94). Around the outside of the W. Kennet barrow came a dry stone wall with upright sarsen stones at intervals. This dry stone walling was a great accomplishment on the part of the builders, and marked an advance. Long-headed skeletons were found in the chamber, and no evidence of cremation. The plan at B is of the Corridor Tomb at New Grange, near Drogheda, Ireland. Externally it consists of a huge heap of stones, 280 feet in diameter and 50 feet high. Internally the corridor is some 60

feet long, and leads to the
central chamber, which is
roughly domed over at a
height of 20 feet. Off
this central chamber are
recesses, used for sepul-
chral purposes. These
chambered barrows are
planned much on the same
lines as the Stone Age
Temples of Malta. Some-
times the bones found in
the Long Barrows are dis-
jointed, as if they had
been placed there some
while after death; and
it may well be that only
the heroes were thought
worthy of such burial.
Because the barrows were

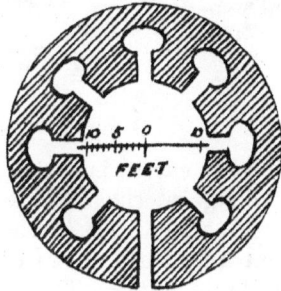

FIG. 34.—Picts House, Sutherland.

used for more than one burial, it has been suggested that
slaves may have been sacrificed to accompany their tribal
chiefs to the spirit world, in the same way that implements and pottery were broken, and animals slaughtered,
but it is doubtful if slavery was yet possible. We shall
probably be quite safe if we regard these barrows as tribal
mausoleums, where the people could assemble and hold services.
They are a visible sign to us that Neolithic man believed in a
life hereafter, and built them as an emphatic assertion that
death is not final. It must have needed some great impulse
to bring the tribe together, and make them willing to under-
take such a vast work as the construction of a barrow.

This provision of houses for the dead throws an interesting
sidelight on the belief of those days; it suggests that in
Neolithic times the spirit was tied to the earth for some
little while, whereas in the later Bronze Age burials, when
the body was burned, it seems as if the spirit was freed at once
to go to the spirit-world (H. G., p. 44). The homes for the dead
may have been modelled on those of living men; there is a
range of habitations which would appear to have been developments of this idea. Figs. 33 and 34 show what are known as
Picts Houses in Scotland, and this form of stone construction

FIG. 35.—Eskimo Rock Hut.

covered with earth is clearly derived from the chambered barrows. Again, the Eskimo houses (Figs. 35 and 36) seem to be survivals carried to the N. In Fig. 35 there is a long tunnel entrance leading to the hut, with the beds at A, and the cooking-places at B. The roof of hut is formed of skins, with a layer of moss between, carried on the poles shown in the sketch. The window is of membrane stretched between whales' jaw bones. The snow house (Fig. 36) is of the same form. There are Picts houses in Scotland which consist of a paved trench lined with masonry, and covered with stone slabs which terminate in a round chamber.

Fig. 37 is of a Picts Tower, Doon, or Broch, found in Sutherland, Caithness, Orkneys, Shetlands, and the Hebrides. The little door shown is only 3 feet 8 inches high, by 3 feet broad, and leads through the wall, which is 10 feet 6 inches thick, with a guard cell off the passage 4 feet high and 9 feet

FIG. 36.—Eskimo Snow House.

FIG. 37.—Picts Tower.

long, with a doorway 2 feet square. There is a circular court inside, open to the sky, and in the wall of this, opposite the entrance, another door leads to a passage winding up in the thickness of the wall to upper galleries, all of which are very low, and lighted by windows into the inner court. It is very difficult to date such buildings, but these Picts towers are Megalithic in character, and built of dry stone ; in design they are first cousins to the Nuraghi of Sardinia, which are fortified dwellings. The Picts are supposed to have descended from the Neolithic stock, and, it may well be, built these towers, perhaps as late as Roman times, in this distant part of the country.

Fig. 38 shows a Cromlech* or Dolmen ; this was part of the chamber of a barrow, from which the encircling earth has been removed, and ploughed away. Its construction is as described on page 38.

Fig. 39 shows a Monolith or Standing Stone, called Maen Hir in Wales, where there are many of them. Probably they mark graves of important persons but they sometimes represent the sole relic of a Stone Circle or of an Alignment. The latter is a double parallel row of standing stones, a feature sometimes (as on Dartmoor) extending for more than a mile. It is usually found in connection with the circle or the round barrow and points to religious ceremonial. The arrangement of one horizontal stone lying across two uprights, as at Stonehenge, is called a Trilithon.

We have said that megalithic means building with giant stones, and it is well to realize how large some of these were.

* The word *cromlech* in the above sense has been in use in Britain since the thirteenth century. Through a misapprehension it has been used by French archæologists to signify the stone circle, and by some English authorities. To avoid confusion the name *dolmen* has been applied lately in both countries.

FIG. 38.—A Cromlech or Dolmen.

Mr. Peet, in *Rough Stone Monuments*, writes of a block weighing nearly 40 tons, which must have been brought 18 miles, at La Perotte, Charente, France.

It may be as well before we pass on to Stonehenge, the greatest of our megalithic monuments, to get some idea of how the builders went to work. It is probable that the only mechanical aid they had was the lever. Boys and girls, who learn mechanics, will not need to be reminded of what the lever means, so they must excuse this digression for some others who may not know.

Fig. 40 shows a see-saw, and the principles of leverage may have been discovered by Neolithic, or perhaps Palæolithic, boys and girls amusing themselves in this way. A see-saw is like a pair of scales; it does not make any difference if you sit on the beam, or are suspended below it. If the two boys sit at an equal distance from the centre, and are of the same weight, they will balance one another, but if one is heavier, he will have to come nearer the centre, if equilibrium is to be maintained. So much is this the case,

FIG. 38.—A Cromlech or Dolmen.

FIG. 39.—A Standing Stone.

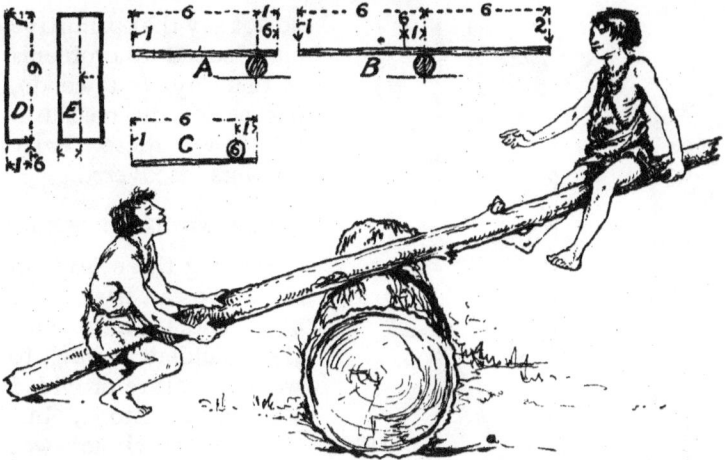

FIG. 40.—The Laws of Leverage.

that if he is very much heavier, say 6 stone, to his small brother, 1 stone, then the heavy boy need only be 1 foot from the centre, to balance the light boy at 6 feet, as A, Fig. 40. Imagine the beam at A as a lever; 1 cwt. applied in a downward direction at one end, 6 feet away from the centre, will exert an upward pressure of 6 cwt. at the other end, 1 foot away from the centre.

If the boys sit, both on one side, as at B, they will be balanced by a 2-stone boy 6 feet away on the other side. If we take the left-hand side of B, and find that 6 stone at 1 foot = 1 stone at 6 feet, and apply it as at C, and imagine the 6 stone at 1 foot as a log or stone which has to be lifted, then 1 stone lift 6 feet away will do it. We can apply our lever in a different way as at D. The beam is bent at right angles; one arm is 6 feet long, and the short one 1 foot. A 1 stone push at the top of the 6 feet long arm will produce a 6 stone pull up at the end of the horizontal arm, 1 foot long. This brings us to the erection of church steeples, chimney shafts, and towers. Take E, 6 units high, by 2 broad in its base, as a tower which has to resist the pressure of wind by its weight. Wind pressures are known, and their force on the whole area is applied to a lever arm of half the height of the tower as at E. To oppose this there is weight, acting through its centre of gravity, on a lever arm of half the width of the base. If the wind pressure is greater than the weight, over goes the tower. We

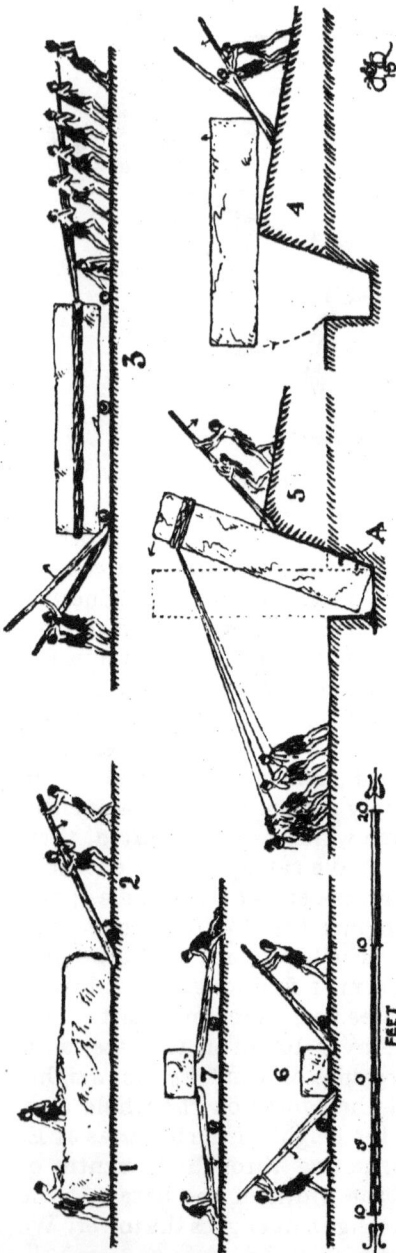

Fig. 41.—Megalithic Builders at Work.

do not say that primitive man looked at problems in this way, but we do, because of the mechanical laws these early builders discovered.

BUILDING STONEHENGE

Bearing these laws in mind, we can pass on to a consideration of how the builders went to work. Nature provided a local sandstone, but the inner circle was constructed of strange stones. The nearest place from which these could have been obtained is the W. of Pembrokeshire, and it may be that the stones were already a sacred circle before being moved. No. 1, Fig. 41, shows the masons dressing the stone into shape in its original position to save weight in transport. It is thought that the masons used fire first to heat the stone, and then water to make fragments split off, but it would be a dangerous method, and they may have used wooden wedges instead. We have seen a good mason in Inverness-shire working on a large granite boulder on the hillside where it was dropped out of the bottom of a glacier ages ago. The mason wanted to make a 6-inch landing, and he obtained this by drilling a series of holes, into which he inserted wedges, and so split the landing out of the heart of the boulder. Neolithic man perhaps used the same methods, but of this we cannot be sure; we do know that he had flint and stone tools, because these have been found when excavating to raise the fallen stones at Stonehenge. The flint axes were roughly sharpened, and held in the hand, and appear to have been used to clean the surface of the stone, after it had been bruised by larger stone boulders, or mauls, which smashed off the bumps.

No. 2, Fig. 41, shows men lifting one end of the block to place rollers under it. No. 3 shows the rollers in position, and men pulling rough hide ropes, with others behind assisting with levers. At 4 we arrive at the building place, where a hole was dug, having one sloping side, and the upright stone being set in the hole, it was fixed by ramming small stones into the triangular space at A 5, but it seems obvious that a sloping embankment as at 4 must have been built up before the stone could be tipped into the hole. Without the embankment it would be nearly impossible to raise the stone, and a very dangerous job. With the embankment, even if the stone slipped forward a little in the tipping over, it could easily be levered back into the hole, and then when resting against the embankment as at 5, pulling and levering would have raised it; meanwhile earth shovelled down into the triangular space at A would have fixed the stone in the desired position. As to the top lintel stones, these may have been placed in position by making a bigger embankment, or by levers as 6 and 7. The stone raised once could be blocked up, and the operation repeated. The stone shown in Fig. 41 is about the size of one of the uprights in the outer circle of Stonehenge. Fig. 42 is a sketch plan showing the original form of Stonehenge. First there is an outer rampart, not shown on the plan, consisting of a circular ditch and bank, about 300 feet in diameter. There is an opening on the N.E. in the circle, where it is joined by an avenue. Within this rampart comes the actual temple as shown on plan. First there is the outer circle at A, which originally consisted of 30 stones, standing about 14 feet high by 7 feet wide by 3½ feet thick. Around on top of these stones comes the circle of

FIG. 42.—Plan of Stonehenge.

crowning lintels, mortised or hollowed out on their undersides on to tenons or stubs worked on the tops of the vertical stones under. Fig. 43 gives some idea of what this outer circle must have looked like when complete. Within this circle is another, at B, of smaller stones, and then at C came 5 magnificent trilithons arranged in horseshoe form on plan. Each trilithon consisted of two upright stones and one lintel, and starting from the N.E., or entrance side, the height of the trilithons is increased. Inside the trilithons is another horseshoe of smaller monoliths at D, around the flat Altar stone at E.

Just inside the entrance from the avenue is a large flat stone, which has the sombre name of the Slaughter Stone, and a little way down the avenue another upright one called the Hele Stone.*

There are many interesting speculations as to the purpose and age of Stonehenge. It will be noticed that it is set out on an axial line which points to the N.E., or where the sun comes up over the horizon on the longest day, or summer *solstice* of 21st June, but it does not appear to do so now on the exact centre line of the entrance avenue, so far as it is possible to determine this. Taking this difference into account, and the astronomical fact that the sun rises each year a little more to the East, Sir Norman Lockyer and Prof. Penrose formed the idea that about four thousand years ago the sun did rise on the actual axial line of the avenue. We have tried to show this in Fig. 44, and have shown the Hele Stone as part of a trilithon. This estimate of age agrees with the archæological evidence, because in the excavations carried out for raising the fallen stones, only flint implements were found, and not any bronze tools which would point to a later date. There is a model in the Prehistoric Room at the British Museum of Sir Norman Lockyer's theory.

As to its uses, it may well be that Stonehenge was a Temple

* Traditionally " Friar's heel." Hele Stone has been substituted

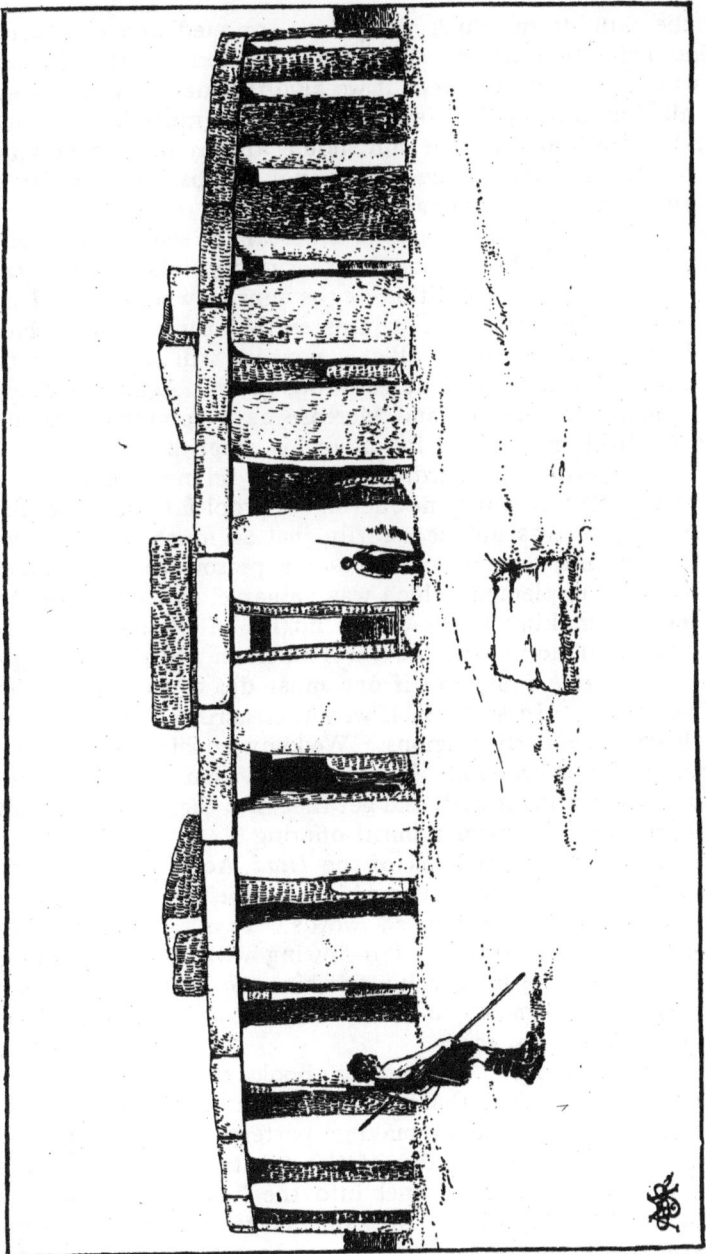

Fig. 43.—Stonehenge.

of the Sun, from which the priests or medicine men could take their observation. We accept the longest and shortest days as a matter of course, if we give the matter any thought at all, but not so the Neolithic man. It must have been a mystery to him, that the sun should appear in a shallow arc across the horizon in the winter, but climbs into the sky in summer-time. It annoys us on dull days to know that the sun shines behind the clouds and we cannot see it, and Stonehenge may have been a magic observatory, where the priests could determine the position of the sunrise when it could not be seen. The priests may have settled the seasons ; have said now is the time to plant ; now we will sacrifice to the Sun-god that he may make our crops grow. Again, we accept the miracle of growth and increase as a commonplace, but the Neolithic man, who, in one of his rough hand-made pots, had safeguarded his hardly won seed, did not commit it to mother earth without some offering, or propitiation, or sacrifice. The sacrifice was not necessarily just so much sheer cruelty as an offering to the gods of some person who was loved, or a pot or implement which was valuable, so that the person or family making the sacrifice might be blessed. The individual did not count for very much in those early days ; the tribe came first, and if one must die to save the others it had to be. In some such way the sacrifice became a part of the ritual of early religions. We know how in Genesis xxii. 2 God said to Abraham, "Take now thy son, thine only son Isaac, whom thou lovest, and get thee into the land of Moriah ; and offer him there for a burnt-offering."

In the twenty-first book of the *Iliad*, Achilles, after he has killed the son of Priam, throws him into the river, and speaking over him "exalting winged words," says, " Nor shall the river avail you anything, fair-flowing with its silver eddies, though long time have you made him sacrifice of many bulls, and thrown down single-hooved horses, still living, into its eddies " (*H. G.*, p. 42).

In Mr. and Mrs. Routledge's book, on the Akikúyu of British East Africa, there is an account of the people who dig for sand for use in making pottery. It is interesting, because it gives us an idea of the spiritual outlook of these people. The natives tunnel into the hillside for sand, like so many rabbits, and as they do not take any precautions, the burrow sooner or later falls in, and smothers the excavator.

NATURE WORSHIP

The Akikúyu do not take any steps to dig the poor fellow out, because this would offend the Spirit of the Sand Pit, but sacrifice a goat instead to propitiate the spirit, then start another burrow which, in its turn, necessitates another goat being sacrificed. This, we think, would have been the case with the Neolithic men : they would worship the Sun, Moon, and Stars, the Rivers and Waters, the Mountains and Valleys, and a great Mother God over all. If by any chance the spirits were offended ; if certain things were done which were taboo, or forbidden, sacrifice had to be made. Just as the Akikúyu appear to be a very kindly pleasant people, who do not enslave one another, or go to war, so we can free the Neolithic men from the charge of cruelty.

Stonehenge does not appear to have had any connection with Druidism, which followed many centuries after. The Druids worshipped the Moon and Stars, and Stonehenge was a Sun Temple, built by an agricultural people, to whom the Sun was all-important.

So far as Neolithic man is concerned, his religion must have been a very real one to him, or he would not have taken so much trouble with the Megalithic monuments we have been describing. These are very widespread, and can be traced along the shores of the Mediterranean, through France, to this country ; we have seen how the Picts towers resemble the Nuraghi of Sardinia (p. 41), and the chambered barrows the Stone Age temples of Malta.

This art of building was in its way as wonderful as the La Madeleine paintings we wrote of on page 94 of Part I., and we must try and imagine the builders. There is a danger in archæology of thinking more of the things than of the people who made them ; we talk of flint implements, as if the New Stone Age could be collected in a bushel basket, and shown in the glass cases of a museum, and especially is this the case in the prehistoric period before there was any written history. The interest of things is that they were made by people, and when the things are temples and tombs they become extraordinarily indicative of the spirit of man ; of that essence, or aura, which gives him and his work individuality, and has made possible the great works of architecture, painting, poetry, and sculpture, and which makes it possible for a man to lay down his life for an idea. Any great movement which appeals to the mind of men has always been compounded of the spirit.

NATURE WORSHIP

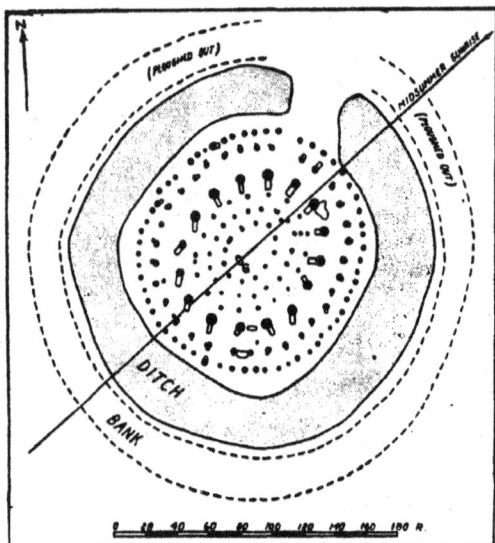

FIG. 44.—Plan of " Woodhenge,"
Salisbury Plain.

We append to this chapter a plan of Woodhenge, the discovery of which has been described in the Introduction (page vi). The name is a regrettable artificial variant of Stonehenge. The site is in the parish of Durrington, two miles from Stonehenge, and it is the site of a group of wooden circles. There are socketed holes of no less than six concentric oval rings of timber posts, now marked by short concrete pillars. They are surrounded by a broad, shallow ditch and a flattened bank beyond. The smaller circle at Arminghall, by Norwich, also discovered by Wing Commander Insull, V.C., has a single horseshoe, once filled by posts twenty or thirty feet high. They both date from the time of the Beaker people.

This plan shows the post-holes (black dots) of the six circles formed by wooden posts at Woodhenge mentioned in the Introduction (p. vi). The " tabs " attached to the larger dots indicate sloping ramps as at A, Fig. 41, cut to facilitate the raising of the posts. The G.P.O. cut similar ramps when raising a telephone-pole. The oblong at G (the centre of the shrine) is a grave in which was found a crouched skeleton facing towards the east with a stone axe-hammer and a beaker (thus establishing a Bronze Age date). Among the finds were two axes made of chalk. These belong to the mystery of the ceremonial. They would have been too soft to use and must have been buried at the outset of the work or the frost would have crumbled them.

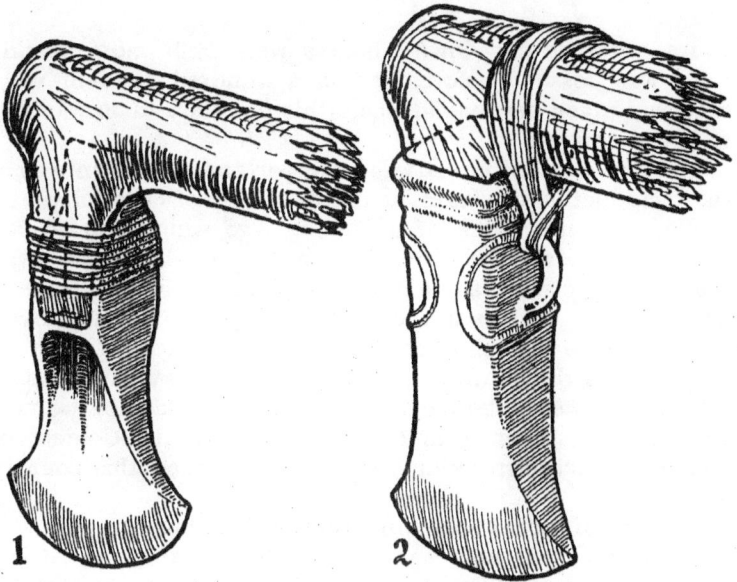

FIG. 45.—Hafting of Palstave and Socketed Celt.

CHAPTER II

THE BRONZE AGE

WE saw, in Part I., how the men of the Old Stone Age found a new material in bone and ivory, and the effect of this was to open up a whole range of new activities. They could make harpoons with barbs in bone, which were not possible with intractable flint. Fishermen should place in their calendar of benefactors the Palæolithic worker in bone who invented the barb.

Even more so the introduction of metal wrought an enormous difference in the lives of men. The Neolithic herdsman, who splintered his stone axe on the skull of a springing wolf, saw the work of months vanish, and was in great danger himself, but when he was the owner of the first of the bronze celts, he walked abroad proudly. The edge of the celt might dull with use, but then it could be hammered up again ; it did not fly into fragments, and it could be hammered cold, which is an important detail to remember.

BRONZE

Trees could be cut down ; houses were built more quickly than was possible before, and in a hundred different ways man was given new confidence in his powers and so was able to make progress.

The discovery of metal was as important in its own way as the introduction of steam, or the discovery of electricity.

We must not think of a Bronze Age which started full blown at a particular date, or of a people who threw away their flint implements one day, to arm themselves with metal on the next. It was a very slow and gradual change over. It is probable that the first flat celts were brought here by traders from the Continent, and many years may have elapsed before they were followed by the round-headed men we now associate with Bronze, and centuries before the Goidels, or first of the Celtic-speaking peoples who reached this country (see p. 12).

The art of Bronze working came from the East, by way of Italy and Gaul, and was widely spread, except in Africa, which never had a Bronze Age. We have seen, on page 12, that the Bronze men were more powerful physically than the Mediterranean race. Probably they were not all armed with bronze, but in any case in the end they conquered the Neolithic people. It was not a conquest of extermination, because we find in the round barrows, which are typical of the Bronze Age, round-headed men side by side with long-headed Mediterranean men.

A parallel can be found in Greece, where the Achæans of the Heroic Age dispossessed the Minoans of Mediterranean stock (*H. G.*, p. 14).

As the art of metal working is the great central fact which has given the name of the Bronze Age to this period, it may be as well to start by a description of the methods followed by prehistoric man in his craft ; in doing so we must try and place ourselves in his position, and imagine that we have never seen metal before. Bronze, we know, is an alloy of copper and tin, and we shall find that copper, like gold, is sometimes found almost pure, and is capable of being hammered up cold, without any preliminary smelting to reduce the ores. Iron ore is found in the form of red earth, or stone, and is not so obviously metallic, and would more easily have escaped attention than copper. The North American Indians hammered up pure copper, and made knives in this

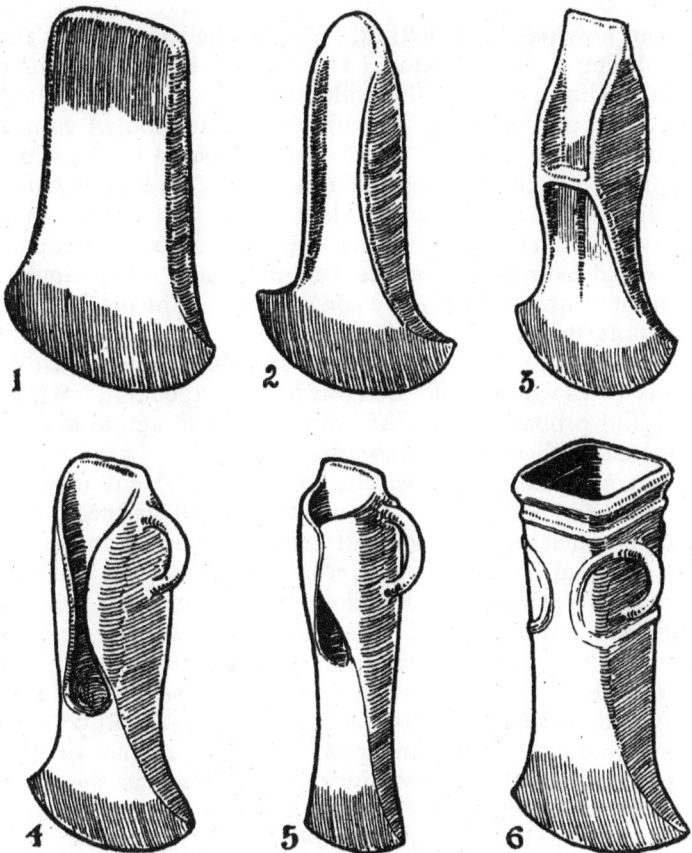

FIG. 46.—Development of Bronze Celt.

way before the coming of the European invaders. So the age of bronze may have been preceded by one of copper. Even when smelting and casting bronze had been discovered, it was found that it could be forged cold, and that when it was heated, it tended to become short and fly to pieces when being hammered. It is hardened by hammering, and softened by heating and quenching, whereas iron hardens by heating and quenching. Bronze was an ideal metal for prehistoric man, because dulled edges could be hammered up again anywhere without very much trouble. It can be made extremely hard.

SMELTING

We can now pass to smelting. Pottery had given man the idea of taking a plastic material and shaping it ; he may have used clay to line a cooking-pit, and found that baking hardened it. In the same way the accidental introduction of copper ore into a cooking-pit, or a charcoal fire exposed to the wind, would have melted the ore, and this would have been found as metal when the ashes were raked aside. The metal may have cast itself into a shape which suggested a tool or weapon, and it would have prompted the ingenious man to experiment. In some such way it must have come about. The first moulds were simple flat open moulds, into which the molten metal could be poured, then progression was made to hollow casting with clay cores which could afterwards be scraped out. Stone, bronze, and probably fine sand were used, and actual moulds can be seen at the British Museum.

We get an inkling of how the bronze men went to work from the *Iliad* xviii. Hephaistos, the famed artificer, who " wrought much cunning work of bronze, brooches and spiral arm-bands, and cups and necklaces," when he starts work on the wonderful shield for Achilles—" went unto his bellows and turned them upon the fire and bade them work. And the bellows, twenty in all, blew on the crucibles, sending deft blasts on every side. . . . And he threw bronze that weareth not into the fire, and tin and precious gold and silver " (*H. G.*, p. 39).

This would have been an apparatus very similar to that used for iron at the Glastonbury lake village, as shown in Fig. 72. Copper melts at 1083° centigrade, and tin at only 232°, so that the Bronze Age founder melted the copper first, then threw charcoal on to the melted mass to retain the heat, and added the tin. The ideal aimed at seems to have been 10 per cent. tin to 90 per cent. copper, but endless experiments went to the discovery that this made a good bronze. Prehistoric man did not know anything about analytical metallurgy. Surface copper ores sometimes contain tin-oxide, and the intelligent man would soon have been moved to find out why a celt made from this ore was tougher than one of pure copper.

We can now pass on to the actual implements made, and Fig. 46 shows the development of the Bronze Celt. No. 1 is called the Flat Celt, and is obviously fashioned on the lines of the stone celt which preceded it, and was hafted in the same way as Fig. 45. The makers soon discovered that

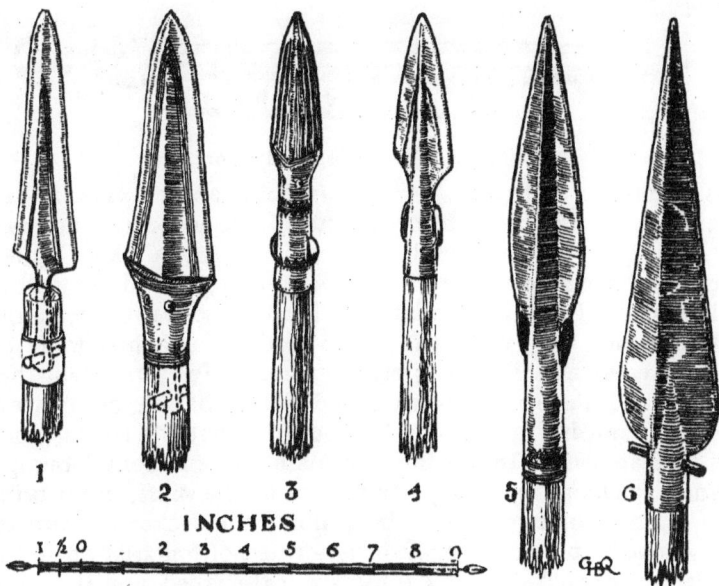

FIG. 47.—Development of Bronze Spear.

by hammering the edge it became thinner, keener, and wider, so the upper part of the later celts is narrower.

No. 2 shows the Flanged Celt, formed by hammering over the sides. This was hafted as 1, Fig. 45. A stick with a stout branch was selected, and this being cut off, was forked to fit over the top of the celt, and bound to it by raw hide. The disadvantage was that the thin celt split the wood head. A stop ridge was then developed between the flanges, and this finally developed into 3, Fig. 46, which is known as a palstave, from an Icelandic word for a narrow spud. This stop ridge took the force of the blow, and prevented the head from splitting (see 1, Fig. 45). In this type, the web between the flanges, above the stop ridge, was thinner than the axe part under, and this feature is more pronounced in 4, where the flanges are hammered over into the form of what is known as the Winged Celt. No. 5 shows the wings lapping, and in 6 they have disappeared, and we arrive at the final Socketed Celt, which was hafted as 2, Fig. 45. There were endless intermediates, and the celt is well worth studying, because it is the ancestor of that friend of man, the axe.

The Bronze Spear is a weapon with an interesting history, It started life as 1, Fig. 47, and in this form was used either

SWORDS AND SPEARS

FIG. 48.—A Leaf-shaped Sword.

as a knife or a dagger. It was cast solid, and provided with a tang which was fitted into the end of the wooden shaft, and this latter was prevented from splitting by a plain bronze collar, through which a rivet passed and secured the end of the tang. In 2 the collar has become socket-shaped, and though not cast with the spear-head, is attached to it by two rivets, and the tang still remains. In 3 the tang has gone, and the socket is part and parcel of the spear-head. But an amusing fact should be noticed: that the rivets which once fastened it to the head remain as ornamental bumps. No. 3 has loops for thong attachment to the shaft, or for tying on feathers or streamers. In 4 and 5 the socket has further developed, and the spear-head is formed of fins cast on to the sides of the socket. In 5 these are leaf-shaped, and the loops are decorative. In 6 the whole spear-head is a triumph of hollow casting.

The Sword developed out of the knife by way of the dagger or rapier. It is easy to see that spear-head No. 1, Fig. 47, if it had a short handle fitted on to the tang instead of the shaft, would make a useful knife. A rapier was an elongated dagger, and the sword a later invention. Fig. 48 shows a beautiful leaf-shaped sword. The tang for handle was cast on the blade, with the edges slightly flanged up, and then in between these edges grips of horn or wood were riveted on each side through the tang, and a round pommel clipped on to the end. Leather scabbards were used with bronze tips called chapes. Bronze was not used for arrow-heads, but flint, as in Neolithic times. The two drawings, Fig. 46 of the celts, and Fig. 47 of the spears, show the development over the whole of the Bronze Age, and by reference to the chart (pp. xi, xii) we shall find that this lasted not less than 1300 years. To realize how long a time this is, we must remember that 1300 years ago in this country would take us back nearly to the time of the death of Ethelbert, king of the Kentish men, and the first English king who received baptism.

These swords and spears are the beginning of the history of Arms and Armour. In the Stone Age men were probably,

FIG. 49.—A Bronze Age Smith.

like the Eskimo, so engaged in finding enough to eat that little time was left for quarrelling about their scanty possessions. With the introduction of metals they could make more things, and life became easier and they had more time for fighting. These Bronze Age arms must have been used in the way described in the *Iliad* (see *H. G.*, p. 26).

FIG. 50.—Bronze
Brooch and Pin.

In Fig. 56 the central man is shown holding a circular Buckler or Shield made of a thin sheet of bronze hammered up into concentric circles of lines and dots. The buckler went with a leaf-shaped spear, as 5, Fig. 47. A flanged celt with slight stop ridge, a type midway between 2 and 3, Fig. 46, was found with a spear-head slightly earlier in form than 3, Fig. 47. The archæologist in this way, by associated finds, builds up a theory of the dates and developments of civilizations. Fig. 49, drawn from the actual tools at the British Museum, shows the equipment of a Bronze Age metal-worker. At 1 are his hammers, hafted like socketed celts. No. 2 shows a tanged chisel, and 3 a socketed gouge. No. 4 is a sandstone rubber, and 5 a quite delightful anvil.

One of the most interesting discoveries ever made in England was what appears to be the complete furnishing of a family at the end of the Bronze Age. This was found in Heathery Burn Cave, County Durham, which may have been used as a house, or as a place of refuge. From remains of skulls which were discovered, the inhabitants appear to have been long-headed men of Mediterranean or Neolithic stock, and it is possible that they removed to the cave in face of the danger of invasion. We shall see later how, at Glastonbury, a people of similar extraction were put to the sword by invaders.

The Heathery Burn discovery included a sword much the same as Fig. 48, but with slight shoulders on the cutting edge of the blade near the handle. A leaf-shaped spear-head, as 5, Fig. 47, but without the loops. Bronze discs 5½ inches diameter, which may have been used as dress ornaments or horse trappings. Bronze collars which fitted on to the nave or hub of chariot wheels, and which, in conjunction with the bridle bit, show that the horse was used. A bucket was found, and tanged and socketed knives ; a razor, gouge, and a socketed celt as 6, Fig. 46 ; chisels, awls, pins, rings, tongs, and gold armlets. There were bone prickers, spindle whorls, skewers, knives the cheek-bars of bridle-bits, and jet armlets ; and all these things can be seen at the British Museum. This splendid find includes nearly all the known types of Bronze Age implements, and we have founded our illustrations on these Heathery Burn discoveries.

The spindle whorl shows that spinning was practised in the Bronze Age in this country ; both spinning and weaving are supposed to have started in the Swiss lake dwellings as early as the Neolithic times. Various types of dress fastenings began to come into use which were suitable for light woven fabrics. Fig. 50 shows a bronze brooch from Ireland, shaped rather like a large hollow curtain-ring, and so arranged that a bronze pin could be passed through it, and in this way fasten a cloak drawn through the ring. This type may have suggested the penannular brooch, as Fig. 76.

Spinning

In a barrow in the East Riding, Yorks, of this period, the remains of a linen winding-sheet were found under a skeleton, and woollen fabrics have been found in others ; these could only have been woven on a loom. We will consider, then, the steps which a Bronze Age weaver had to take if she wished to convert a fleece into a piece of stuff for making clothes. It would need washing and cleansing first, and then came dyeing. Crotal, a lichen growing on trees, may have been used. If this is put in a pot with the fleece and water, and boiled for one or two hours, it produces a rich red-brown colour. Teasing consists of pulling the fleece into fluff, and oiling explains itself. Carding is an operation which consists of putting the wool on an implement rather like a large butter-pat with teeth, called the card, and then pulling the other card across it, so as to arrange the wool for spinning. This latter was the occupation of girls for so many centuries, that we still talk of an unmarried woman as a spinster.

The spindle which was used in the Bronze Age consisted of a piece of wood, perhaps about 1 foot long and $\frac{1}{2}$ inch diameter, and a few inches from one end came the whorl, which acted as a miniature fly-wheel and helped to twist the spindle. At the other end was a little nick in which the yarn was fastened. In spinning, a roll of carded wool was held in the left hand, or bound on to a distaff ; from this roll a little wool was pulled out and twisted by the fingers until a piece of yarn was made about 18 inches long, and this was tied to the spindle. The wool was then paid out with the left hand, and the spindle twisted with the right. When the spindle stopped revolving it was held, when the twist ran up the length of wool which had been paid out and made this into yarn, which could then be wound on to the spindle and the spinning resumed. We have shown this in Fig. 51.

WEAVING

Weaving is, and has been since the Bronze Age, one of the crafts which has had the greatest influence on the progress of man. It is beautiful work, done wherever man wants clothes, and carried out in many different ways; but the main principle of weaving is always the same. The long threads running through the length of a piece of cloth are called the warp; the ones which cross these by going under and over the warp are called the weft. From the discovery of loom weights, as shown at the bottom of the warp-threads in Fig. 52, in the Swiss lake villages and in England, it is thought that the earliest looms were of this pattern, which is called the Warp-weighted Loom; the weights keeping the warp properly stretched. The warp-threads are kept in place by yarn threaded through them at the bottom. It is probable that at first the weaver took the skein of yarn in

FIG. 51.—Spinning.

her right hand, and picking up the warp-threads one or two at a time with the left hand, passed the weft-threads through from side to side, over and under the warp. She may have used a wooden lath to beat the weft-threads up, and so make the cloth compact.

Fig. 53 shows the next development, and our drawing is based on the Scandinavian loom in the Copenhagen Museum. The diagrams at the side, A and B, illustrate the method of weaving, and we shall find as we go along that, though the details are elaborated, this principle remains. A piece

FIG. 52.—Warp-weighted Loom of Simplest Type.

of fabric has been woven at the top downwards, and below this the warp-strings hang down with their weights on the ends. They are divided at 1 by a shed-stick : the shed is the space through which the weft is passed. At 2 is the heddle-rod, which is attached to alternate warp-strings by loops. The weaver then passes his shuttle through the space between the warp-strings, above the heddle-rod in A position, which is called the counter shed. The heddle-rod is then pulled out to B position, which brings the warp-threads which were at the back to the front, and the weft is again passed through the space now called the shed.

In this way the weaving proceeds, like darning, first under and over the warp-strings, then over and under. This would make a plain cloth ; in patterned work different coloured yarns can be used, and instead of just over and under the

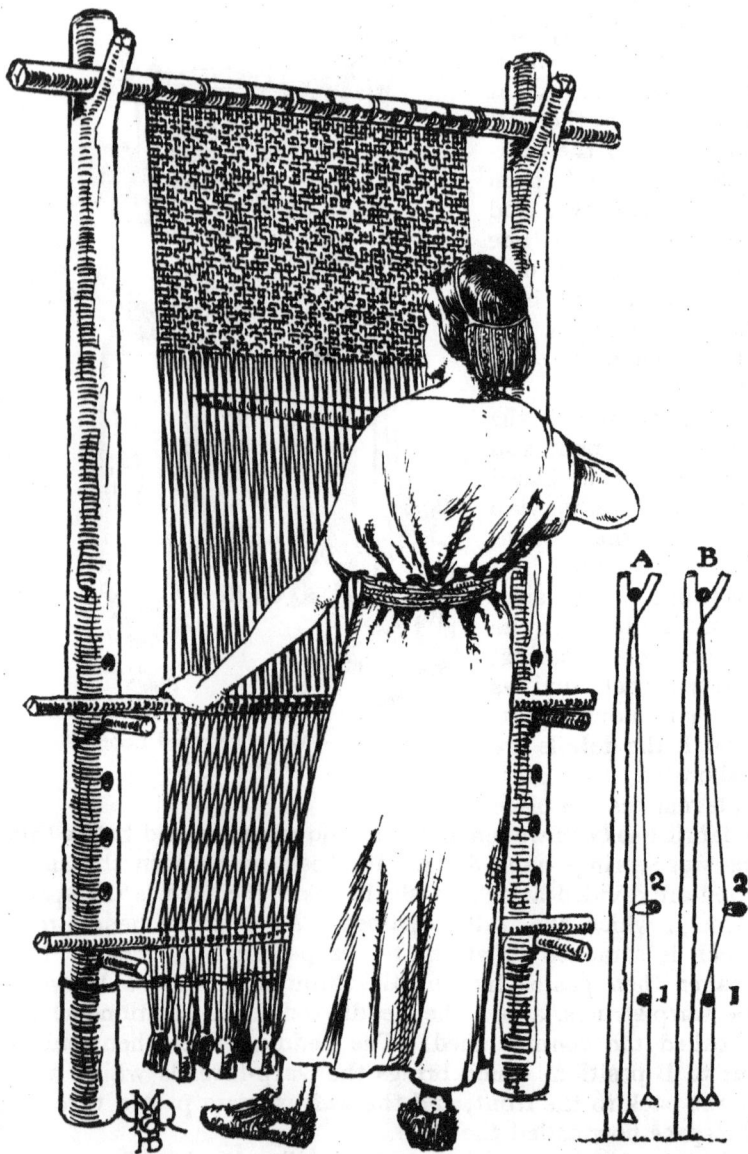

FIG. 53.—Warp-weighted Loom of more Developed Type.

warp, you can go over and under and then skip two or three, and so produce a pattern. On Greek vases Penelope is shown working at an upright warp-weighted loom like Fig. 53, but it has been developed by making the top cloth beam to revolve, so that the cloth could be wound up as it is woven (*H. G.*, p. 75).

Fig. 54 shows what is called now a weaver's comb, found at Glastonbury lake village, but we doubt if this was used, as suggested, to comb or pack the weft-threads tightly together; it would have been an inconvenient way of doing it; so here is a problem for our readers to determine the use of the comb.

Fig. 55 shows a man shaving with a razor of a very usual pattern in England during the Bronze Age; he probably used oil instead of soap.

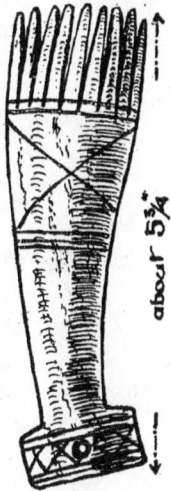

FIG. 54.—Comb.

Fig. 56 is a costume plate for the three periods of this book, and it is the central figures which are of Bronze Age and so discussed here. The remains of dresses of this period have been found in Jutland, which suggest that the piece of stuff woven on the looms was wrapped around the body without any shaping. This is the case with the tunic of the man and the skirt of the girl. In the case of the man this was the beginning of the kilt. The girl's bodice would have been roughly cut in kimono shape, and the side seams sewn under the arms. She is shown wearing a bronze disc fastened on to a woven tasselled belt, and her hair was gathered into a thread net, and fastened by long bronze pins. She is wearing a jet necklace. The shoes of both man and woman are of skin, and the man has a circular cloak and cap of thick rough knotted wool.

We have seen on page 58 that one of the finds at the Heathery Burn Cave was a point of deer antler, about 5 inches long and curved in shape; it is pierced twice on the radial lines of the curve, and once at right angles. Similar pieces have been found in the Swiss lake dwellings, and it is suggested that these were the cheek bars of bridle-bits, as Fig. 57. Probably the first bit was a twisted leather thong, knotted at the width of the mouth, and then the ends passed through the cheek-pieces as reins. If

FIG. 55.—Shaving with Bronze Razor.

the transverse hole of one of these horn bars is examined, it will be found to be worn smooth as by a leather rein. Similar cheek-pieces are described in the *Iliad* (*H. G.*, p. 26). A sketch is added to the drawing of a bronze bit from the Swiss lake dwellings, which shows the influence of the early antler type. The pony in Fig. 57 is wearing the gold Peytrel, or breast-plate, discovered at Mold, Flintshire, which is now in the British Museum. It would fit a pony of about twelve hands, and it is embossed in the same style as the bucklers. When one bears in mind that the warrior to whom it belonged did not in all probability decorate his horse, until he had satisfied his own vanity, we can be quite sure that together they must have presented a splendid sight.

The Heathery Burn discovery includes bronze nave collars for chariot wheels. The nave of a wheel is its hub, and this suggests spokes. The first wheels were probably solid on their axle, rather like a cotton reel. A, Fig. 58, shows another type made up of three boards secured by dovetailed clamps. B, Fig. 58, shows the start of the spoke, not as we know it to-day, but arranged more as a brace. The upright part includes nave, two spokes, and parts of the felly or rim, all in one piece of wood. The four other spokes are braced between this and the remaining parts of the felly. These come from the Swiss lake dwellings, and must be early types, because a later wheel has been found there which, though in bronze,

To face p. 64.]

New Stone

Fig. 56.—Costume of the
Bronze

and Early Iron Ages.

BRONZE

FIG. 57.—Bridle and Gold Peytrel.

65

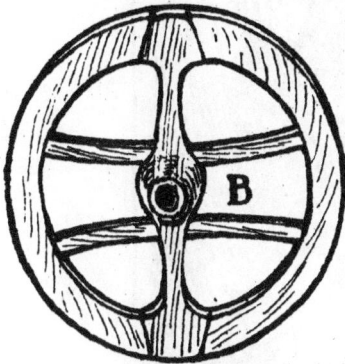

2' 10" diam.

2' 0" diam.

FIG. 58.—Wooden Wheels.

must have been founded on a wooden construction. It is 19¾ inches in diameter, and has four spokes radiating between nave and fellies, just like the wheel of to-day. We know too that beautifully turned wooden wheel naves have been found at Glastonbury lake village, dating from the Early Iron Age, and in what are called the chariot burials of Yorkshire, of the same period, the iron tyres of chariot wheels have been discovered.

The original Aryan-speaking peoples, the forerunners of the Celts, are supposed to have possessed ox-wagons, and it may well be that chariots were introduced by the Goidels, who reached these shores from 700 to 500 B.C.

The chariot does not give very much opportunity to the maker to vary its shape. There must be a floor framed up on the axle, around which would come the body, perhaps of wickerwork covered with hides. There would have been a centre pole, with yoke attachment to the horses. The chariot of classical times must have been founded on some such simple basis as Fig. 1. Again we cannot do better than turn to the *Iliad* for an idea of how chariots were used (*H. G.*, p. 116).

This question of wheel naves, the discovery of jet armlets at Heathery Burn Cave, and shale cups in round barrows, all of which must have been turned, brings up the question of lathes. It is difficult to see how a simpler turning contrivance than the Pole Lathe (Fig. 78) could be made, and this may date from the Bronze Age.

Fig. 59.—A Clapper Bridge.

COMMUNICATIONS

FIG. 60.—Plough.

The Hill Forts and Camps were still the rallying places of
the people, and it is probable that places like Badbury, Maiden
Castle, and many others which had been started by the Neolithic men were improved upon in the Bronze Age. The
trackways on the hilltops between the camps would have
become more defined as traffic and trade routes, with tumuli
to mark the line. Fords may have been replaced by bridges ;
there are two on Dartmoor which are still called Celtic. Fig.
59 shows one of these at Postbridge, and its construction is
just what we should expect from a people who had inherited
the building tradition of Stonehenge. We should like to
draw attention to the trumpet shown in the hands of one of
the figures. These instruments derive their shape from the
horns of animals, which had been used for the same purpose
before. They were made at the end of the Bronze Age, in that
metal, and are supposed to have been used by the Celtic people
in warfare ; of two types, some have the mouthpiece at the
side.

The possession of the bronze celt, with its better cutting
powers, meant that man could make ever larger clearings in
the forest, grow more corn, and keep more herds. He was
helped again, because with his bronze sickle the harvesting of
his crops was not such a problem as when that useful implement was of flint, as Fig. 21. There is a beautiful harvest
scene in the eighteenth book of the *Iliad*—" where hinds were

reaping with sharp sickles in their hands. Some armfuls along the swathe were falling in rows to the earth, while others the sheaf-binders were binding in twisted bands of straw. Three sheaf-binders stood over them, while behind boys gathering corn and bearing it in their arms gave it constantly to the binders ; and among them the king in silence was standing at the swathe with his staff, rejoicing in his heart. And henchmen apart beneath an oak were making ready a feast, and preparing a great ox they had sacrificed ; while the women were strewing much white barley to be a supper for the hinds " (*H. G.*, p. 40). Game was less eaten now than the domesticated animals ; a proof that life was becoming easier, and it was not necessary to live by the chase. There are Scandinavian and Ligurian rock carvings of Bronze Age date, which show a primitive plough drawn by oxen, as Fig. 60, but England was the very outpost of civilization in those days, and we cannot be sure that the plough reached here so early ; yet it would not have needed so much cleverness to make as a bronze celt, once the idea became known.

The hut of the hut circle was much the same as in Neolithic times, built in the Berm of the camp or just around it ; but from remains which have been found, it looks as if the hut itself was becoming less pit-like, and rising out of the ground with vertical side walls, as Fig. 67. It must be remembered that the Bronze Age men had their enemy the wolf, waiting always just round the corner to cut off stragglers, so we may be sure they lived in communities.

Pottery was still hand-made, without a wheel, but ornament was improving, and consisted of straight lines arranged as chevrons, lozenges, herring-bones, with dots and concentric circles, as Fig. 61. No. 1 in Fig. 62 is a Beaker, or drinking-vessel, which was introduced on the East Coast by the Beaker people, see page 11 ; it is found with unburnt burials. No. 2 is a Food Vessel. No. 3 a Cinerary Urn, made to hold the ashes of a cremated burial ; and No. 4 an Incense Cup. This does not mean that the Bronze Age people used incense, and the name has been suggested by the pierced treatment of the little cups ; these are found in Round Barrows, and may have been used to bring the sacred fire which started the funeral pyre. It is thought that these types of pottery, which were doubtless deposited with the dead, for their use in the spirit world, are similar to those they used in their everyday life. Bronze

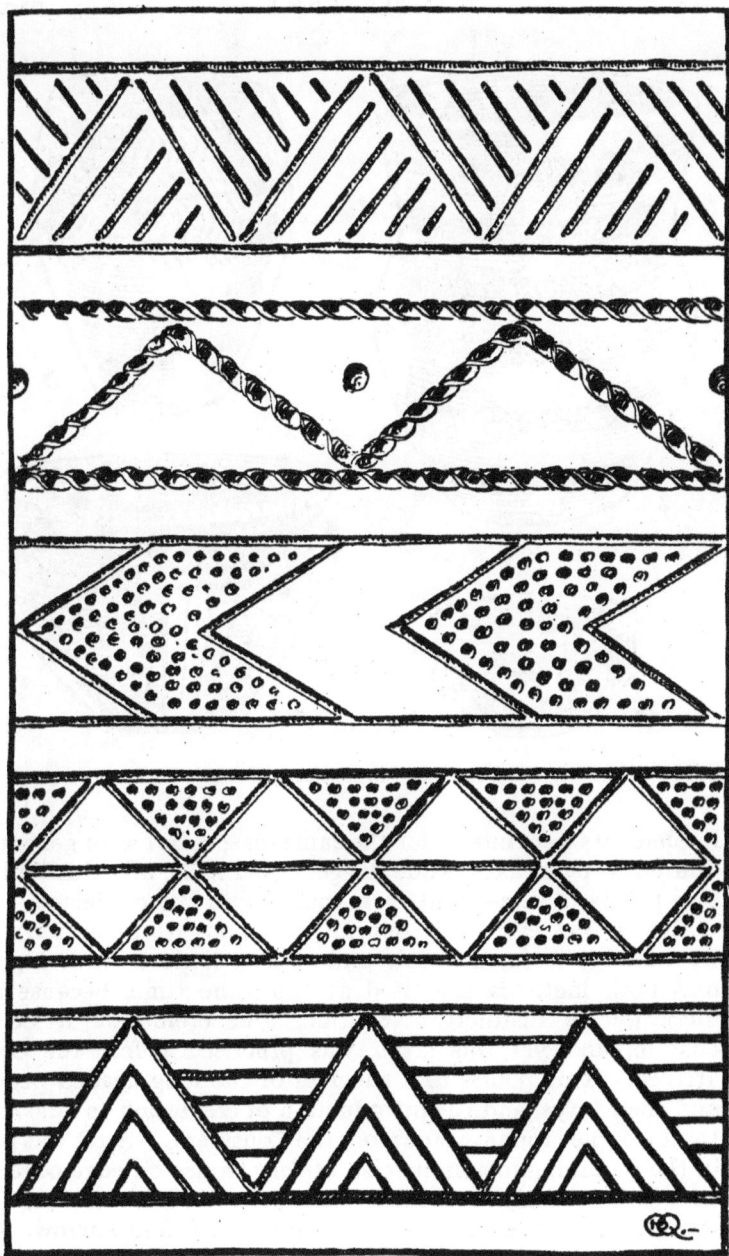

FIG. 61.—Bronze Age Ornament.

70

FIG. 62.—Bronze Age Pottery.

implements were buried for the same reason, but were generally limited to plain axes, knife daggers, and awls, and this limitation points to some symbolical meaning in those selected.

Burial was either by burying the body (inhumation), or by burning it (cremation), and it is a little bewildering to find both methods practised at the same time, because inhumation is distinctly Neolithic, and cremation a Celtic custom, and yet this latter was practised before the Celts arrived. This points to a survival of the long-headed people and their ways, and the introduction of cremation as a fashion by the earlier round-heads from the Continent. A pit was dug in the ground, and a stone cist, of four stones on edge covered by another, made to cover it, or a hole cut in the chalk, and the ground heaped over in the form of a round barrow. In a stone country, the barrow was made of heaped stones, and became a cairn. No. 1, Fig. 63, is the type which is called a

Bowl Barrow, because it is like an inverted bowl. No. 2 a Bell Barrow, because the ditch and bank made around the outside give it that shape; and No. 3 is a Disc Barrow.

A barrow is sometimes called a Tumulus; in Derbyshire, a Low; and in Yorkshire, a Howe.

Silbury Hill, 6 miles W. of Marlborough, on the Bath Road, is in the form of a round barrow, but it is 135 feet high, and covers 6 acres. It is wholly artificial, and in 1907, at the rates of pay then obtaining, its cost was estimated at £20,000.

Cup and ring markings are common on the cover stones of the cists or graves in the barrows, and these are very similar to the markings found on the churingas of the Australian aborigines (p. 64, Part I.).

Small objects called Sun Discs are found in Ireland; these are made of gold about $2\frac{3}{4}$ inches diameter, and have the same decorative idea as the cup and ring markings, made up of concentric circles. All these things point to Sun-worship being characteristic of the Bronze Age; another symbol, which is widely distributed, is the swastika, also considered a symbol of the Sun.

It must be borne in mind that prehistoric man was still held in thrall by magic and mystery; that there were many things which were taboo or forbidden; like the Akikúyu his life and death were governed by a complicated ritual. Cremation in all probability was not practised to destroy the body, but to purify it of sins and uncleanness, and render the spirit fit for the life hereafter. In the twenty-third book of the *Iliad* the spirit of the hapless Patroklos appears to Achilles and urges him : " Thou sleepest and hast forgotten me, Achilles. While I lived never did'st thou forget me, and only now that I am dead. Bury me with all despatch, so that I may pass the gate of Hades. Far do the spirits keep me off, the spirits of men out-worn ; they suffer not that I should join their company beyond the River ; and vain are my wanderings through the wide-gated house of Hades. Pitifully I beg that thou should'st give me thy hand ; never again shall I come back from Hades, once you have granted me my due of fire " (*H. G.*, p. 44). We have seen that the implements which were buried with Bronze Age man were limited to certain symbolical types. Again we find that in the actual cinerary urns were buried, with the human remains, the bones of wild animals, like the fox, mole, and mouse ;

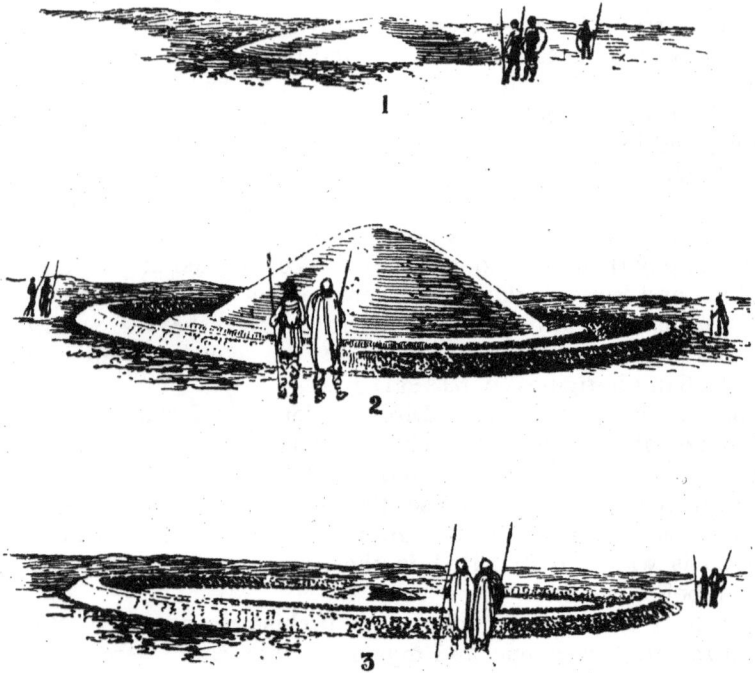

FIG. 63.—Bronze Age Barrows.

surely these typified something. In the barrow itself, the bones of the ox, goat, sheep, horse, pig, and dog have been found with cremated burials ; of these some may be the remains of the funeral feasts, and the horse and dog may have been slaughtered to accompany their master, and the sacrifice of slaves and captives may have formed part of the ceremony. Bone pins have been found, charred by fire, as if they had fastened the body in its shroud before it was burned.

Homer, in the twenty-fourth book of the *Iliad*, gives a wonderful picture of the burial of Hector :

" So nine days they gathered great store of wood. But when the tenth morn rose with light for men, then bare they forth brave Hector, weeping tears, and on a lofty pyre they laid the dead man, and thereon cast fire.

" But when the young dawn shone forth, rosy-fingered Morning, then gathered the people round glorious Hector's pyre. Assembling, they first of all quenched the flames of

73

F

HECTOR

the pyre with wine, even as far as the might of the flames had reached, and thereupon his brethren and friends gathered his white bones, mourning him with big tears coursing down their cheeks. The bones they took and laid away in a golden urn, wrapping them up in soft purple robes, and quickly set the urn in a hollow grave, and heaped above great stones, closely placed. Then hastily they piled a barrow, while everywhere about watchers were posted, through fear that the well-greaved Achaians might make an onslaught before the time. And, when the barrow was piled, they went back and, assembling, duly feasted and well in the palace of Priam, that king fostered by Zeus. Thus did they hold funeral for Hector, tamer of horses."

Even fuller details are given in the twenty-third Book of the funeral of Patroklos, and the funeral games. Of how they went forth " to hew high-foliaged oaks with the long-edged bronze," and " splitting them asunder the Achaians bound them behind mules," and so brought the wood to the appointed place, and made a great pile. " And they heaped all the corpse with their hair that they cut off and threw thereon." The pyre was " a hundred feet this way and that, and on the pyre's top set the corpse." " And many lusty sheep and shambling crook-horned oxen they flayed and made ready before the pyre; and taking from all of them, the fat, great-hearted Achilles wrapped the corpse therein from head to foot, and heaped the flayed bodies round. And he set therein two-handled jars of honey and oil, leaning them against the bier; and four strong-necked horses he threw swiftly on the pyre, and groaned aloud. Nine house-dogs had the dead chief : of them did Achilles slay twain and threw them on the pyre. And twelve valiant sons of great-hearted Trojans he slew with the sword " to be consumed by the fire. The North Wind and the loud West " all night drave they the flame of the pyre together, blowing shrill," and after a barrow was made as already described for the burial of Hector. Then followed the funeral games, of which all can read in the twenty-third book of the *Iliad* (*H. G.*, p. 46). The next time we see a Round Barrow, we must think of it, not as only so much heaped earth, but rather as a visible sign of our own Heroic Age. We must try and conjure up a picture of the flaming pyre, and looking across the smoking eddies of time, see the crowd of Bronze Age warriors burying their chief.

Now we think we had better try and give our readers some idea of the migrations and minglings, the traffic and trade routes, which had developed in the Bronze Age from the earlier Neolithic beginnings. We must first ask ourselves, why it is we find these big movements of men, because, leaving on one side the adventurous few, the general run of people do not move until they are pushed. In the Old Stone Age, man moved because he was a hunter, and had to follow the chase to live, and in the same way, even when he had settled down, he could not be sure of a permanent home, unless it was accompanied by a perennial food supply ; if this failed, then he had to break fresh ground. If food was one of the reasons for his moving, he naturally went away from the crowded central area, or falling on his neighbours compelled them to do so. Wars have played a terrible part in migrations ; we shall see in our time great movements of people, as a result of the 1939–1945 struggle. The study of these movements is of great value as bearing on the original homes of men. That is why the archæologists continually do dig ; they are hunting for first causes.

Geography will help us to discover the natural causes of man's movements on certain lines. On p. 14, Part I., we referred to the Loess land. Loess is a sandy, chalky loam, deposited at first as dust blown by great blizzards from the glaciers in the Ice Ages. This loess is in a broad zon , which, starting from the Ural Mountains, stretches across South Russia to the Carpathians, and the Danube, then through North-West Austria to South Germany, and the North of France. It is shown by dots on Fig. 64. The fine grain of the loess prevented the spread of forests, and became instead the great grasslands which have played so considerable a part in the development of Europe. Here have been bred great hordes of men, who in times of drought, or when the regions became overpopulated, have descended on the ancient civilization of the East, and caused movements of men. In the same way, the Arabian Desert has been a great reservoir of hardy people, who periodically have made exodus, with terrible happenings to their prosperous neighbours, or have been bribed to keep the peace.

The problem which confronts such a people is similar to that of the hill-tribes of the N.W. frontier of India. Here the Mohmands, Afridis, Wazirs, and Mahsuds, perched on the barren hills, can only live by levying tribute on the caravans passing from the fat lands. Here through the great land gate of the Kyber Pass, through all the ages, immigrants have gone into India. The Aryans, and Alexander : all travelled on this line until we forced a new way by sea.

If along a certain line similar kinds of pottery or stone monuments are found, it is fair to assume that these are the work of a particular type of people moving along this line. If in Bronze Age barrows, we find gold from Ireland, glass or beads from the Mediterranean, amber from Scandinavia, or in an Early Iron Age cemetery at Aylesford in Kent, a bronze flagon from North Italy, it points to trade and trade routes. We may be sure that salt was traded.

We have already written, on page 10, of one of the earliest migrations, that of the Mediterranean people ; on page 11, of the first of the round-heads ; on page 12, of the arrival of the Beaker people ; and, on page 12, of the movements of the Aryan-speaking peoples. This brings up another factor of great importance in the lives of men, and one which is not concerned so much with their movement, as with the circulation of some great idea that acted as a lever, and caused them to alter their mode of living. The wonderful drawings and paintings of the Aurignac and La Madeleine periods, in the Old Stone Age, which we discussed in Part I., and the Megalithic buildings of the New Stone Age, were wrought around some central inspiration ; again, in the latter half of the Bronze Age, the prophets were at work, and we find the introduction, by the Aryan-speaking peoples, of cremation and all that it may have implied. The Minoan civilization was centred in the island of Crete, the home of Minos, and then transferred to Mycenæ on the mainland of Greece. The Cretans were of the Mediterranean stock ; and if reference is made to the chart on pages xi, xii, it will be seen that their power declined about the 1500 century B.C. Their buildings were megalithic, and they did not cremate their dead. While the Minoan civilization was dying, we hear of the beginnings of the Heroic period of the Hellenes. Jason, Agamemnon, Hector, and Odysseus are typical of wild men who came from the N., finding their way down from the grasslands shown on Fig. 64, and they were an Aryan-speaking people who cremated their dead (see H. G., p. 14). The Achæans were followed by the Dorians, who destroyed the Mycenaen civilization in Greece,

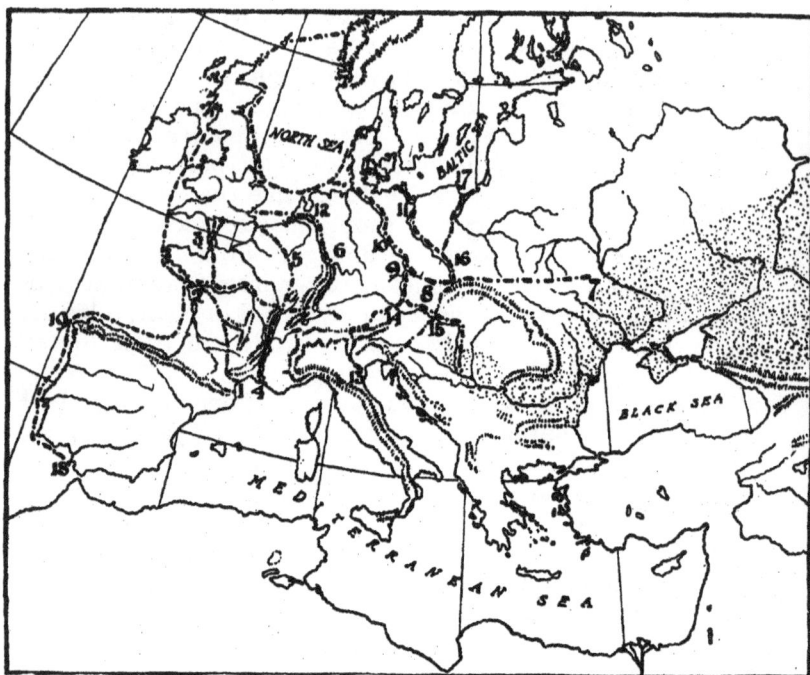

FIG. 64.—Traffic and Trade Routes.

and settling down became the Spartans. There were great movements of the Celts, Gaels, or Gauls, in the Early Iron Age. They were a Nordic people living to the north of the Alps, and called by the Romans for this reason Transalpine. They sacked Rome 395 B.C., and were typical of the barbarians across the Danube and Rhine who were to become a constant menace to the Empire later on.

If the Mediterranean men found their way through Gaul, on a line 1, 2, 3, Fig. 64, a later route seems to have been from Marseilles (Massilia) at 4, by the Rhone Valley to Châlons, where it divided into three lines ; one to the W. down the Loire to 2, the second around the Paris basin at 5, and the third through the Belfort Gap, between the Vosges and Jura Mountains, and down the Rhine at 6. This latter route is an important one, because it mingled people coming up from the Mediterranean, with another type coming from the regions to the N. of the European and Asiatic Mountains.

MIGRATIONS

Prof. Fleure, in his paper on the *Racial History of the British People*, thinks that the Beaker people came from Kiev on the Dnieper, S. of the Pinsk Marshes (7, Fig. 64), and in Mr. Crawford's paper on the Bronze Age Settlements, we find a map of the localities in which their distinctive pottery has been found; at 8, on the tributaries of the March in Moravia; on the Bohemian tributaries of the Elbe by Prague; around the junction of the Saale and Elbe at 10; the mouth of the Oder at 11; on the Zuyder Zee at 12; and again at the junction of the Rhine and Main at 6. Mr. Crawford shows how pottery beakers of the same type are found on our eastern coasts from Caithness to Kent, and also found on the W. coast of Scotland.

The W. coast of Denmark, and the S. Baltic, supplied amber during the Bronze Age, and the B.M. Guide Book for that period gives the two main trade routes through Germany to the Adriatic. One started from Venice at 13, Fig. 64, up the valley of the Adige, through the Brenner Pass, down the Inn to Passau on the Danube, at 14, and then by way of the Moldau to the Elbe, and so by the line 9, 10 to Denmark. The second route was from Trieste to Laibach and Graz, then to Pressburg on the Danube (15, Fig. 64), from there up the River March, across Moravia and through Silesia, along the Oder at 16, then across Posen to the Vistula, and Dantzig at 17. The spiral design of the Bronze Age found in Scotland, Cumberland, Lancashire, Northumberland, S. Ireland, and Merionethshire, and which was common in Egyptian and Ægean art, is supposed to have found its way here on the first of these two routes.

We can now pass from land journeys to sea voyages, and we will work back from Cæsar's time. It was the Veneti, maritime tribesmen occupying what is now Vannes Morbihan, in Brittany, who formed a confederation of the tribes in N. and N.W. Gaul against the Romans. The Veneti seems to have controlled the trade with Britain, and possessed a fleet of large ships with leathern sails, high poops, and towers, but did not use oars, which was the reason they were beaten on a calm day by the Romans.

If we go back again to the time of Pytheas of Marseilles, about 330 B.C., we find that he sailed to Britain, and there was in his time a regular trade between Cornwall and Marseilles, and probably a sea-borne trade between Cornwall and Cadiz

78

(Gades) (18, Fig. 64), which was a centre of the tin trade. From Cape Finisterre, Pytheas sailed E. along the N. of Spain to Corbilo at 2, on the mouth of the Loire, past Ushant to Land's End (Belerium), where he landed. He sailed all round Britain, and attempted an estimate of its circumference, and indicated the position of Ireland. Long before this, as we have just seen, the Beaker people came across the North Sea, and settled on our East Coast ; so even the prehistoric period had its great seamen and sea-faring traditions.

This enables us to take up the question of the position of the Cassiterides (from the Greek word for tin, *cassiteros*), or the tin islands of the ancients : were they really islands ? The Greeks and Romans obtained tin from Galicia (19, Fig. 64), Cornwall, and possibly the Scillies, but the main supply was from Cornwall, and possibly it is the British Isles which were the Cassiterides.

Pytheas says tin was conveyed by the people of Belerium in wagons, at low tide from the mainland, to the island of Ictis, where it was purchased by merchants, carried to Gaul, and transported on pack-horses to Marseilles, the overland journey taking thirty days. To start with there has been considerable doubt as to the locality of Ictis ; some think it was S. Michael's Mount, others the Isle of Wight or Thanet. The tin must have been mined in Cornwall, and it would have meant a long overland journey to the two latter places.

We have seen there were good sailors, and the general weight of evidence inclines us to accept Dr. Rice Holmes' view, that the tin was shipped at S. Michael's Mount, close to where it was mined. The fact that the Veneti formed the confederation against Cæsar points to a predominance based on trade, and they may have controlled the tin traffic, in which case Corbilo (2, Fig. 64) would have been a natural place for unshipment.

From Corbilo to Marseilles is approximately 500 miles on 2, 1, 4 line, which means nearly 17 miles a day for the pack-horses on the thirty days' journey. The tin was cast into ingots, of the shape of ankle bones, and 2 = load for a pack-horse.

Britain has always been rich in metals. Copper is found in Cornwall, Cardiganshire, Anglesey, Snowdonia, and in Ireland. Tin in Cornwall and on Dartmoor. Prehistoric man would have obtained his copper from boulders, or found

TRACKWAYS

lumps of ore on the hillside, and tin from the gravel beds of streams. Ireland was El Dorado of the Old World, and gold was found in the Wicklow Hills as late as 1795. It was shipped across to Carnarvonshire, or the mouth of the Mersey, and from there found its way down by way of Shrewsbury, Craven Arms, Wootton Bassett, Sarum, and a deeper and more navigable Avon to Christchurch, and so across to Cherbourg. Another route appears to have been from the Mersey, across the Peak District to Peterborough and the Wash, where it was shipped to Denmark and North Germany.

Mr. Crawford's paper on *Early Bronze Age Settlements* is an interesting illustration of how, by mapping the finds of bronze implements, and gold ornaments, trade routes are established. Sea-borne traffic is shown by the large number of hoards of bronze implements, found near the seacoast, and around the estuaries of navigable rivers. Déchelette proved the same thing in France.

Going right back to Neolithic or perhaps Palæolithic days, we find that flints were mined at Grime's Graves (Grime=the devil) in Brandon and at Cissbury near Worthing, and apparently only roughly chipped there and then exported to be finished elsewhere. They must have been carried along the trackways to the hill forts. These old trackways have interesting names. The Ridgeway comes from Fenland along the Dunstable Downs to Berkshire, the White Horse, and the Marlborough Downs ; there is the Harroway coming from Cornwall, and finding its way through Hampshire to the Thames estuary ; and the Pilgrims' Way, along the southern slopes of the North Downs, was an old road long before men tramped its surface to Becket's shrine at Canterbury.

Here we must attempt to sum up what we have found out about the Bronze Age. The introduction of metal opened up new activities for man, and especially new opportunities for the individual. The Neolithic man toiled with antler pick and shoulder-blade shovel, and piled earth in the banked camps. He chipped sarsen stones, and fidgeted them into the upright position of menhirs and dolmens. It was patient team work in which every one laboured for the community. He needs must move from camp to camp to find pasture for his flocks. In much the same way primitive peoples like the Tasmanians, Australian aborigines, and the Eskimo are fully occupied in hunting to live ; they have not any leisure for fighting, or any possessions to fight for. When everything has to be carried about, the lighter you travel the better.

The earlier round-heads appear to have been powerful, and may have been a pleasant people; we have seen that they were buried side by side in the same barrows with the older stock of Neolithic long-heads, and this points to friendly conditions. These early round-heads carried on the building traditions of the New Stone Age; the hill camps were improved, and they may have had some hand in the completion of Stonehenge, but hardly a trace of bronze has been found there.

As metal became more plentiful, larger clearings were made in the forests, and man began to settle down. He could grow more crops and keep more cattle; he began to have possessions. This was the opportunity for the individual; if a man was harder working than his fellows or more far-seeing, cleverer or more frugal, he could become a man of property, and, founding a family, become the chieftain. The tribe was gradually forged into a nation, and the chieftain became a petty king.

We may be sure that this wider life brought in its train a set of problems which had not confronted the Neolithic herdsmen. As man began to have more possessions, he became alarmed for the safety of his own, or envious of those of others. The elaborate planning of the later hill forts points to the necessity for being prepared to withstand raids, and it may be that we must look to the Bronze Age for the beginnings of organized warfare.

A people who could plan earth-banks in so subtle a fashion as the entrances of Maiden Castle, Dorchester, give proof of being able to work together, and so may have attempted, in a gradual way, to solve the problem of the right mode of living. Without some code or tradition, the community of a hill fort would have degenerated into a rabble. We shall find as we go along that man is tremendously concerned with this, and seeks many ways for his own government. We shall not be far wrong if we picture the Bronze Age people as living, like the Homeric Greeks, under kings and nobles, yet given some share in the framing of the law.

CHAPTER III

THE EARLY IRON AGE

HERE we must start by another reminder : that at the beginning of the Early Iron Age, which first saw the introduction of that metal, men did not pack up all their old bronze implements and bury them in hoards, to at once arm themselves with iron. It was, on the contrary, a very gradual change over, and for a long time both bronze and iron were used side by side. This was so at Hallstatt in the Noric Alps of the Austrian Tyrol. Here there have been salt mines from the earliest times, and it must have been an important trading centre. Excavations have been carried out in the cemetery of the salt miners, and the implements found there have been held to be distinctive of the civilization at the beginning of the Early Iron Age, when bronze was still in use.

The second half of the Early Iron Age is held to be most typically shown by implements which have been recovered from an old settlement, built on piles, on the margin of a bay on Lake Neuchâtel, near Marin, to which the name of La Tène or the Shallows, has been given. The finest developments of the Early Iron Age are to be found in this country, since it fell under Rome's influence at a later date than the Continent ; in the same way the Iron Age, or Late Celtic tradition, survived in Ireland and parts of Scotland which were never occupied by the Romans.

The people of England had become very mixed racially. On page 10 we sketched the order of the arrivals of the different peoples ; and just as bronze overlapped the use of iron, so the old peoples carried on their everyday life and were not always exterminated by the new-comers or even dispossessed of their lands. We saw how, in the early Round Barrows, the later round-heads were buried side by side with the earlier long-headed Iberians.

The next arrivals were the Goidels, or first of the Celtic-speaking peoples. On page 14 we mentioned the generally accepted theory that they were driven into the W. by their successors, the Brythons, who were related to them and spoke another variety of the Celtic language. This is now being given up, and it is thought that there were never any Goidels

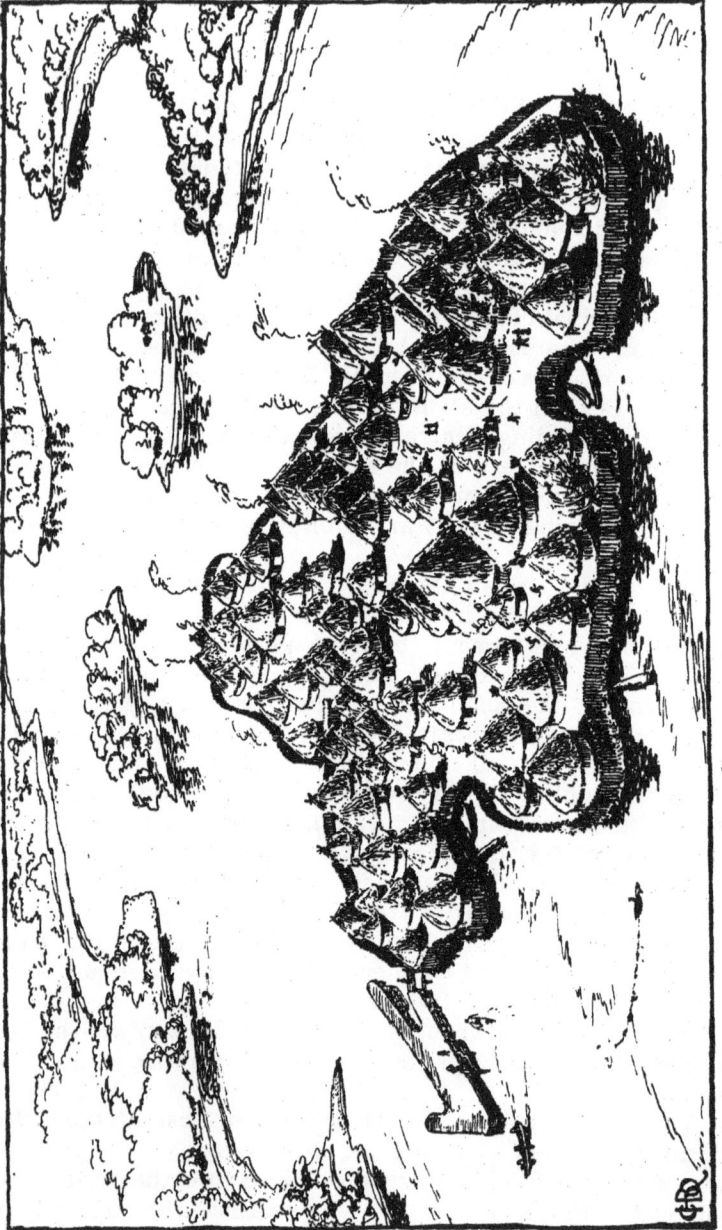

FIG. 65.—Glastonbury Lake Village.

LAKE DWELLINGS

in England or Wales, but that they went directly to Ireland, the Isle of Man, and Scotland, where their Celtic descendants still live.

The Brythons were followed by the Belgæ, who, while they were responsible for the finest developments of what we now call Late Celtic art, were not themselves of pure Celtic stock. They came from where Belgium now is, and had more Nordic blood than their predecessors ; they were a half-Teutonic and fierce fighting people.

We saw on page 58 how the people of the Heathery Burn Cave were of long-headed stock, which yet had absorbed a Bronze civilization. Much the same thing occurs in the Iron Age at Glastonbury lake village, and we shall base our illustrations of the period on the houses and implements discovered there.

On page 103, Part I., we referred to the Azilian dwellings, built over the water. In Neolithic times this idea was developed, and in Switzerland there were dwellings built on the margins of lakes. They were first discovered at Ober-Meilen, Lake Zürich, in 1853, and this started research, and the discovery of similar structures in different parts of Europe. These may be divided into three types. (1) The Swiss dwellings, built on platforms formed on the tops of piles driven into the lake bed, which date from the Neolithic and Bronze Ages and resembled in form those which are now built by the inhabitants of New Guinea. (2) Another type in which, instead of pile foundations, large open framings resembling log huts were sunk in the lake and steadied by piles, much like the modern caisson used by engineers for foundations. Dwellings of this type were built in France and Germany during the Early Iron Age. (3) The type like Glastonbury and the Scottish and Irish Crannogs. These were really small islands formed in the middle of marshes, and being stockaded around, were raised above the flood-level by earth brought from outside ; but the foundation was quaking bog, which, as we shall see at Glastonbury, gave the inhabitants a great deal of trouble. These date from the Early Iron Age, and continued to be occupied in remote spots, as places of refuge, until the seventeenth century.

As the Swiss lakes became overpopulated, people moved downhill into the Po valley, and here are found the settlements which are called Terremare, from *terra marna*, or marl

FIG. 66.—Hut Interior at Glastonbury.

GLASTONBURY

earth. The peasants discovered that the earth from these old settlements was valuable for agricultural purposes, and in carting it away came across antiquities which disclosed the secret.

There are literary references to lake dwellings. Cæsar said, writing of the Morini (a Belgic tribe in Gaul, opposite Kent): " They had no place to which they might retreat, on account of the drying up of their marshes (which they had availed themselves of as a place of refuge the preceding year), and almost all fell into the power of Labienus " (*Com.* iv. c. 38).

Venice itself, the Queen of the Adriatic, is a glorified crannog which started as a place of refuge. " They little thought, who first drove the stakes into the sand, and strewed the ocean reeds for their rest, that their children were to be the princes of that ocean, and their palaces its pride."

Hereward the Wake maintained himself, in the last stand against the Norman, in the marshy recesses of the Isle of Ely.

Now we come to the interesting way by which we in England came to be provided with a lake village of our own. Mr. Arthur Bulleid of Glastonbury, when he was a young man, read Keller's *Swiss Lake Dwellings*, and was fired with the idea that there must have been a lake village in the olden days in the swamps near Glastonbury. Remember that in this neighbourhood there is the tradition of Arthur and his knights and the Isle of Avalon :

> " The island valley of Avilion,
>   Where falls not hail or rain, or any snow,
>   Nor ever wind blows loudly."

So whenever Mr. Bulleid went on his walks abroad he kept a wide-open eye for any indications of a possible site for a lake village. This was in the end discovered by the mounds which had been left where the hut foundations were, and though in the course of 2000 years or more the land had been drained, and became covered with vegetable soil and turf, yet these mounds were still noticeable to the observant eye. In the molehills were found pieces of bone and charcoal, and when Mr. Bulleid made a trial hole he came across more charcoal, some pottery, and two oak beams. Again, a labouring man, David Cox by name, told Mr. Bulleid that when he had been cleaning out a ditch about three-quarters of a mile away, in 1884, he had found a black oak beam embedded in the

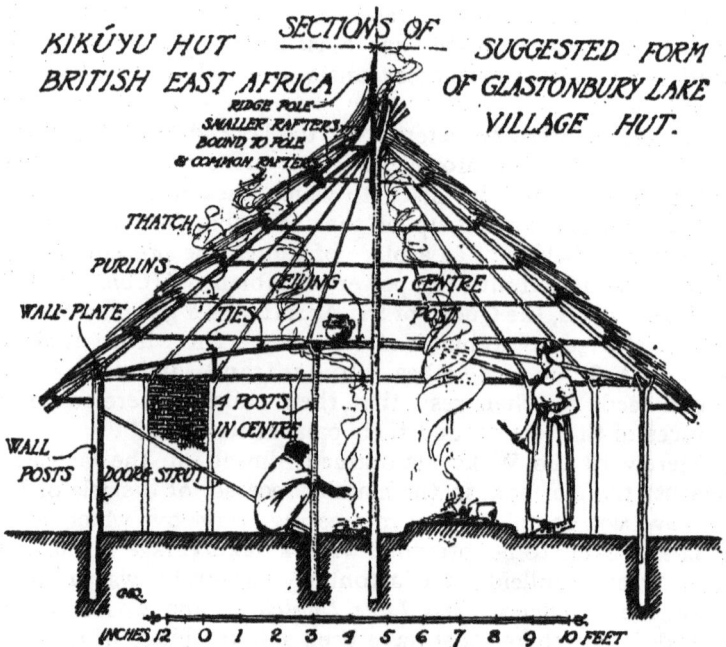

KIKÚYU HUT SECTIONS OF SUGGESTED FORM
BRITISH EAST AFRICA OF GLASTONBURY LAKE
RIDGE POLE VILLAGE HUT.
SMALLER RAFTERS
BOUND TO POLE
& COMMON RAFTERS
THATCH
PURLINS
WALL-PLATE CEILING 1 CENTRE
TIES POST
WALL 4 POSTS
POSTS IN CENTRE
DOOR STRUT

INCHES 12 0 1 2 3 4 5 6 7 8 9 10 FEET

FIG. 67.—Hut Sections.

soil, and had to cut a piece off it to widen the ditch. Cox
reported that this beam looked like the end of a boat, and this
is what it turned out to be, and it is shown in Fig. 69. So
Mr. Bulleid's dream had come true, and he had found his
lake village. Excavations were started in 1892, since when
the village has been thoroughly explored, and in 1911 a
splendidly detailed account was published in book form by
Mr. Arthur Bulleid and Mr. Harold St. George Gray. Boys
and girls should endeavour to see these volumes, which are
models of how such work should be done.

Fig. 65 gives a bird's-eye view of the village. The area
was about 10,530 square yards, and the foundations of the
enclosed space were reinforced with layers of logs, laid down
crossways, and filled in with brushwood, stones, and clay,
but it could never have been what the land agents describe
as a " desirable building site." During the time that Glastonbury was occupied, a bed of peat accumulated in some places
5 feet thick, and the inhabitants were constantly rebuilding.

S. PAUL'S CATHEDRAL

The village was palisaded around, with piles driven into the peat, and filled in with wattle and daub. This method was also used in the construction of the huts—there were 80 to 90 of these, roughly circular in shape, and varying from 18 feet to 28 feet in diameter; they may not all have been houses; some were probably used as barns or workshops. The huts contained a central hearth, as Fig. 66, of flat stones let into a clay bed, and as many as 9 or 10 hearths have been found added one on the top of the other, as the foundations settled down into the bog. The wattled walls of the huts were daubed with clay; this is known because pieces of clay showing the marks of the wattles were discovered in the excavations. Each hut had a central pole or roof tree, than this we can gather little more.

We have to look to a primitive people, then, to find parallel building traditions. The Akikúyu, of the Kikúyu hill country, in British East Africa, build to-day, and live in houses which must be the living spit of those at Glastonbury. Fig. 67 shows these on the left hand-side of the section, and on the right is the suggested form of the Glastonbury hut. We have made this drawing from the plan and carefully detailed particulars in Mr. and Mrs. Routledge's book, *With a Prehistoric People*. It is an interesting fact that the constructional problem which the Akikúyu have to face, when they build their huts, is similar to the one which confronted Wren when he designed the dome of S. Paul's Cathedral.

We have seen how Neolithic man built little houses with rafters leaning against a central pole, and this was a very sound method. So long as the feet of the rafters were firmly fixed into the soil, the house stood firm, in gales and under a load of snow; the drawback was that there was no headroom around the walls, and so one had to sit there as you do now in a bell-tent. A wall was raised around to give headroom, as Fig. 19, and this was satisfactory so long as the wall was built of stones heavy enough to provide a sufficient abutment for the thrust of the rafters. The trouble came when the same idea was attempted with thin wooden walls, which would have been overturned.

The Akikúyu first set up about 19 forked posts in holes dug in a circle of about 15 feet diameter. To appreciate the cleverness of the construction, you must remember that none of the wood is bigger than a man's arm. Four posts

DAVID COX

are set up on an oblong in the centre about 4 feet 6 inches by 3 feet. Around the tops of the outer posts, long pliant rods are woven, and these form the wall plate, and take the thrust of the roof. Again ties are woven from this wall plate across from side to side, picking up the tops of the centre posts on the way. Wren took up the thrust of the brick cone which supports the dome and lantern at S. Paul's, by an iron ship's cable, which was let into the stone, and run in with molten lead. We think the rest of the construction of the Akikúyu hut is explained by the drawing.

At Glastonbury there were also found remains of an earlier type of hut, built with wall plates resting on the tops of piles driven into the peat. The huts were apparently oblong in shape, with hurdled walls mortised to the wall plates. Of these we cannot attempt any reconstruction, but of the circular huts we can be more sure, and it seems fair to assume, from what we know, that they resembled those of the Akikúyu.

This building in wattle and daub continued as a tradition throughout Celtic Britain. William of Malmesbury, writing in the twelfth century A.D., mentions the " Ealde Chirche," the ancient church of S. Mary of Glastonbury, built in the seventh century of wattlework.

We know that the Glastonbury people used canoes, for one was found by David Cox, to which reference has been made, and some form of canoes would have been absolutely necessary to the inhabitants of the village. Judged by the peat deposit, all this district around the river Brue must have been a vast morass in the olden days, and in times of flood an inland sea. The canoe (Fig. 69) is of the greatest interest, about 18 feet long; the flat bottom is 2 feet wide, 10 feet from the prow, and its maximum depth inside is 12 inches. It is becoming boat-like, and shows a notable development on Fig. 6, having a shapeable prow, and a graceful rise, or sheer, at bow and stern. The lake villagers had a landing-stage and dock attached to their home, with vertical walls made of stout grooved oak planks driven into the peat, into which were fitted horizontal boards, as Fig. 69. We know they went fishing, because lead net sinkers have been found. Their canoes would have been used to take them to their cornfields on the mainland, the island village had no room for these. Fig. 70 shows a piece of timber found at Glastonbury, and shaped in such a way that it is thought it

FIG. 60.—Dug-out Canoe and Landing-stage at Glastonbury

FIG. 70.—Ploughing.

may have been used as a hand plough, but we are very doubtful of its suitability for this. Many querns and millstones have been found : the earlier type as Fig. 22, and the later rotary types as Fig. 71. In these the lower stone was fixed, and had a wooden pivot in the centre. The top stone was fitted over this, and corn fed through the hole, made large enough to allow it, passed down, and was ground between the upper and lower millstones, coming out at the sides as flour. Small cakes were found at Glastonbury, made of unground wheat grains which had been mixed probably with honey and baked.

FIG. 71.—Grinding Corn.

The villagers also owned horses ; many harness fittings have been found, bits, and the wheels of chariots. Whether the horses were transported to the mainland on rafts or stabled there we cannot be sure. In the summer they may have been pastured on the mainland, within the protection of a camp, and in the winter ferried across to the village to share the huts with the inhabitants. The people doubtless

FIG. 72.—Smelting Iron.

used their canoes to carry on trade with the surplus goods which they manufactured and wished to exchange for other commodities. The two iron currency bars found point to this (see p. 112).

When we pass to the life carried on within the village, we have proof of many and varied activities, but it will perhaps be well to start by a description of the iron working, which gives the period its name.

Fireclay crucibles have been found at Glastonbury, and funnels (*tuyère*) for conducting the blast into the furnace, but it is thought that the crucibles were used for melting copper and tin, to make bronze, as described on page 54.

So far as iron working was concerned, it is probable that this was carried out as the present-day smelting operations of the Akikúyu of British East Africa, which we have shown in Fig. 72. The iron ore is collected from surface workings in the form of an iron sand; this is washed to get rid of the clay and other substances, so that the iron grains are left. The furnace consists of a kidney-shaped hole in the ground lined with clay. The ore is placed in the pit of the furnace, and a charcoal fire started, then more ore and charcoal are added as needed. The blast is introduced at one end of the furnace, which is slightly lower than the middle, by means of a fireclay funnel (*tuyère*). In the funnel are introduced the wooden pipes of the bellows, which are in this way protected

FIG. 73.—Saw and Adze.

from the fire. Two bellows are used, of goats' skins sewn into the shapes of rough cones, or fools' caps, the pipes being

FIG. 74.—An Iron Knife.

attached to the small ends. At the larger ends of the bellows, which are open, are fitted two short sticks, sewn to the skins, but leaving one-third of the circumference free. The smiths' boy holding the two sticks of the two bellows, two in each hand, opens first one bellow, as if the sticks were hinged at one end, and then the other, and closing the opening by shutting his hand, depresses the sticks, and kneads the ends of the bellows, sending forward a continuous blast into the furnace. This blast raises the temperature of the furnace, just as a fire is brightened up by ordinary bellows.

The ore is reduced to a sticky mass rather than molten metal; furnaces which will generate a sufficient heat to make the metal flow, only date from the seventeenth century, and we do not find any cast iron before then. The lump of iron is left in the furnace overnight to cool, and then turned out in

FIG. 75.—Bronze Finger-ring.

the morning, and broken up into sizeable pieces which are forged up into ingots or blooms. This iron is very pure, and ductile, and so can be readily forged; being smelted with charcoal it is free from the sulphur which comes from coal when it is used, and which makes the iron short and brittle. The fireclay crucibles we have referred to were buried in a hole in the ground, and the fire and blast arranged as in the case of the iron smelting.

In Messrs. Bulleid and Gray's book are shown illustrations of all the finds in the excavations, and here we can see daggers, spear-heads, swords, knives, bill-hooks, sickles, saws, gouges, adzes, files, bolts, nails, rivets, keys, and bits. The weapons are few and far between, and this is perhaps one of the reasons the villagers fell an easy prey to their enemies in the end. The man in Fig. 66 is holding an iron bill-hook in his hand, of a quite modern shape; and Fig. 73 shows one man using a curiously shaped saw, with the teeth arranged so that it cuts on the up-stroke, while the other has an adze, which is first cousin to the axe. Fig. 74 shows a man using a particularly beautiful iron knife found at Glastonbury.

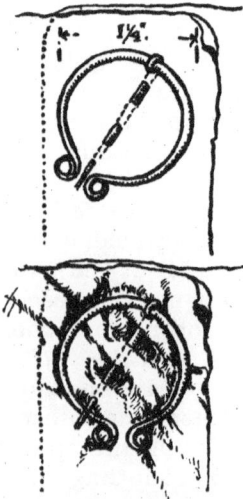

FIG. 76.—Penannular Brooch.

Leaving iron working, we can turn to bronze, which still continued in use in the Early Iron Age as it does to-day.

Fig. 75 shows a bronze finger-ring, and Fig. 76 a penannular (almost a ring) brooch. The top drawing shows how

BROOCHES

the pin, which was loose on the ring, was pushed through the material, and then fastened by moving the ring round a little, and clipping it under the pin. This form of brooch was the forerunner of the buckle.

Fig. 77 shows three bronze brooches, or *fibulæ*. These fastenings came into use in the Swiss and Italian lake villages when cloth was first woven. The three examples drawn here, show the development of these pretty little things, which the archæologists associate with the lake village of La Tène, on the lake of Neuchâtel, and are called types 1, 2, and 3, though only type 2 occurs at La Tène itself. In No. 1 the foot is bent back until it touches the bow of the brooch. In No. 2 the end is no longer free but actually attached to the bow, and in No. 3 the foot and bow are designed as one.

On the right-hand side of Fig. 77 we have drawn the development of the springs of these brooches, and in each case the pin of the brooch is shown vertically. In those of Hallstatt the springs are on one side of the head; those of La Tène are bilateral. No. 1 shows the earliest type, like that of a safety-pin of to-day; so our old friend is of ancient descent. No. 2 has a double coil; and in 3 the pin has one coil to the right, and the wire is then carried to the left, where, after a treble coil, it swings up to form the bow of the brooch. In 4 there is a double coil on one side, and in 5 a treble coil, but the tension is increased by the ingenious way in which the loop or chord across is taken under the arch of the bow; the whole pin—coils, loop, and bow of the brooch—being in one unbroken length. In 6 we have pin and coils to the right, the loop or chord and the coils on the left in one piece; but the bow is a separate part which is hooked under the chord. No. 8 is on the same principle, but the spring is covered with a metal sheath attached to the bow. In 7 the bow is fixed on to a smaller loop. We consider these springs of the greatest importance: 1 dates from perhaps as early as 1400 B.C., and 8 takes us up to the Roman occupation, and, so far as we know, 1 is the first application of the spring. The old brooch-maker who, in 1400 B.C., tapped his bronze wire around a rod and discovered the spring, would have been rather surprised if he could have looked into the future and seen the many ways to which his invention would be applied; for example, that we should tell the time by little spring-driven machines, which we call watches.

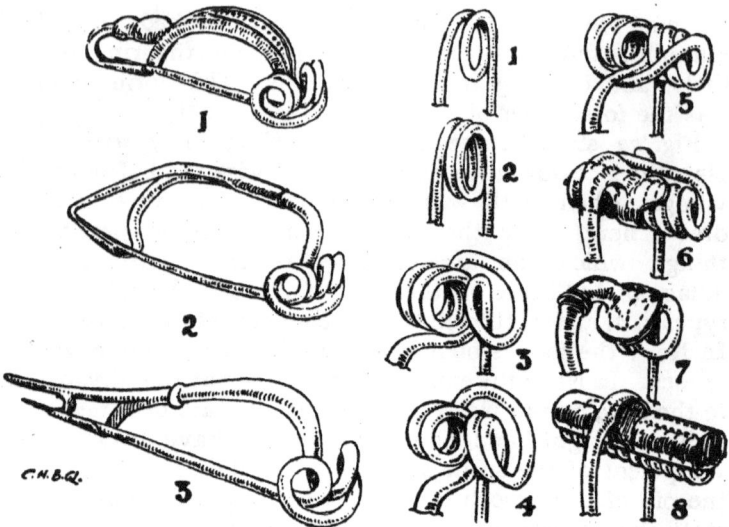

FIG. 77.—Brooches and Brooch Springs.

There were excellent potters at Glastonbury, and Fig. 66 shows some of the pottery found there. The greater part of it appears to have been hand-made, as described on page 26, but the very beautiful pot in the foreground has been turned on some sort of wheel. We saw (p. 25) how the Akikúyu build up their pots on a pad of leaves, which makes it possible to turn the pot round as it is being made, and it is probable that the potter's wheel was preceded by a turn-table, on the lines of the rotary quern (Fig. 71). If a heavy block of stone or wood were pivoted in this way, its weight would aid the momentum of its spin and be very helpful in making pottery. This early type is suggested at A, Fig. 87.

SPINNING AND WEAVING

Spinning and weaving were carried on in the village, and the spindle whorls and loom weights suggest that this work was done as already described on page 60.

TURNING

There were expert coopers at Glastonbury, who knew how to build up tubs with wooden staves and hoops. They were good turners. There is a turned bowl, shown in the lower right-hand corner of Fig. 66, which was decorated in addition with a beautiful running pattern cut in an incised line. There is no evidence of what the Glastonbury lathe was like, but Fig. 78 shows a very primitive type in use in the Chilterns, called the Pole Lathe. It is difficult to see how anything could be simpler than this, and it is obviously a development from the Bow-drill shown on page 73, Part I. In the Chilterns the men who make chair legs buy a fall of beech in the woods, and to save cartage build themselves little huts and turn the chair legs there. The supports for the lathe are often two trees growing close together, which they cut down at a height suitable for the two planks forming the bed of the lathe, into which the poppet heads are fixed. A third sapling is bent down, and the cord, which is to transmit the power, is fastened to this, passed around the chair leg, and connected to the treadle under. A rough tool-rest is provided. The turning is done on the down stroke, which revolves the chair leg towards the turner, and when he takes the pressure off the treadle, the pole pulls it up again ready for another cut. The work proceeds very rapidly, and we have seen chair legs turned, one in a minute.

In our sketch we have shown the turner making a wooden bowl, like the ones which were used before the days of enamelled iron. The block of wood was placed directly against one centre of the lathe, and on the other side came a circular piece of wood, around which the cord was passed ; this was put on to the other centre of the lathe and fixed to the block for the bowl by four brads. This, we think, shows that the so-called Kimmeridge Coal-money is the core left from turning shale bracelets on pole lathes. Coal-money is found near the Kimmeridge shale beds on the Dorset coast, and consists of circular discs, having a hole on one side, and a square recess or two or three smaller holes on the other. The diagram at the bottom of Fig. 78 shows how we think a shale bracelet was turned on a pole lathe. AA are the poppet heads, and BB the centres, C is the circular piece of wood around which the cord was passed, fitted on to one centre, and let into one side of the piece of shale, in a square recess, or by two or three separate pins. The shale being in contact with the other centre. The turner trued up his bracelet, and set its outside shape first, and then making a

FIG. 78.—A Pole Lathe.

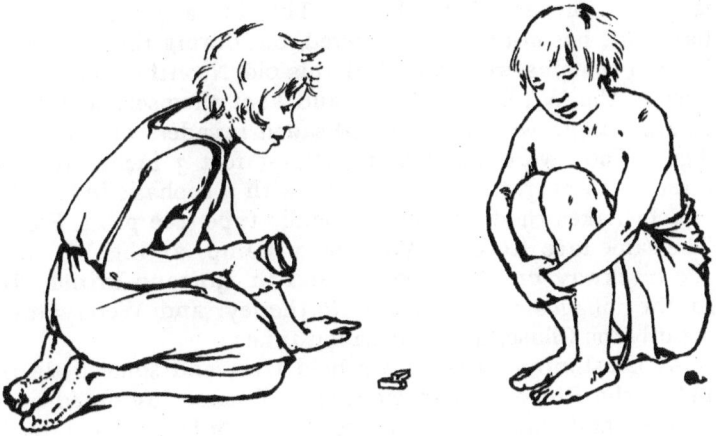

FIG. 79.—Dice.

cut on each face, finally detached it as dotted line D, and the Kimmeridge coal-money was the useless core, and never used as money. One great advantage of these old pole lathes was that the turner could make two or three bowls in graduating sizes from the same block of wood.

The Glastonbury carpenters used axes, and we do not realize in these days what a useful tool this can be, that is, if you are a craftsman and not a wood butcher. Alex. Beazeley, a pleasant architect, and most architects are pleasant, wrote in 1882, that the Swedish carpenters at Dalcarlia and Norrland, " require no other tools than the axe and the auger, and despise the saw and plane as contemptible innovations, fit only for those unskilful in the handling of the nobler instruments : they will trim and square a log forty feet long as true as if it had been cut in the sawmill, and will dress it to a face that cannot be distinguished from planed work." As we jog along we shall find the truth of this, that so long as man is master of his tools we get good work, but when the machine masters the man we have indifferent results.

Fig. 79 shows that there were bad boys at Glastonbury, or perhaps men, who gambled with dice.

The form of lake villages suggests that they were built by timorous people, living in fear of fiercer neighbours. They appear to have had their beginnings with the long-headed Mediterranean race of the New Stone Age. This is borne out at Glastonbury. The burial-place of the inhabitants has not been discovered, but during the excavations human remains were found of this old Neolithic type, which, here in the W., had lived on, and kept themselves free from intermarriage with the round-headed invaders of the Bronze Age. They were small and dark—5 feet 3 inches to 5 feet 8 inches in height. Oval-headed, with a cephalic index of 76, which makes them of mesaticephalic type (see p. 24, Part I.). The same race lived at Worlebury Camp, at the W. end of the trackway on the Mendips, and in Romano-British times in the villages of Woodcuts, Rotherley, and Woodyates, in Cranborne Chase, down to Saxon times.

At Glastonbury their fears held true, and some little time before the Roman occupation, final disaster descended on the village, and they were put to the sword: perhaps by the Belgic invaders, who were long-heads, but of an altogether tougher fighting breed. Cæsar (*Com.* v. c. 43) tells us how the Nervii, when attacking Cicero's camp, set fire to the thatch of the huts, by discharging redhot clay sling bullets. Many of these were found at Glastonbury, and help us to visualize the final scene. We have noted that very few arms were found in the excavations, and the little dark men only wanted to be left quietly alone, and be allowed to get on with their work; and this is what they did until they were discovered. Then their outlying possessions and crops would have been destroyed and the village surrounded. The Glastonbury men could only have watched the scene, in shuddering misery, from behind their stockades, and then the invaders, using perhaps the dug-outs they had collected from the waterside, would have paddled across the lake, and discharging their redhot clay bullets have fired the thatch. When the flames subsided, the few survivors would have been put to the sword. Yet the little dark men have had their revenge; from the very start of their career they appear to have lived in communities; it may have been a tradition they brought with them from the shores of the Mediterranean. The Belgæ who oppressed them, like the later Anglo-Saxons, whom they resembled, preferred a more open-air life, and to-day their fair-haired descendants have the same tastes.

Prof. Fleure, in his paper on the *Racial History of the British People*, sums up the matter thus : " These descendants of the Neolithic people are the long-headed, long-faced, dark-haired,

Fig. 80.—Coracles.

FIG. 81.—Framework of Umiak.

brown-eyed people that form so strong an element of the
population of big English cities. They seem better able than
all other types to withstand slum conditions, so that in the
second generation of great city life they have arisen in their
millions to form once more, after many days, almost a
majority, perhaps, of the population of S. Britain." So the
tale of the Mediterranean men is not yet completed.

We have seen how fond the ancient Britons were of wattlework, and on page 90 how it was used even for the construction
of churches. Boats were made in this way, and Fig. 80 shows
a coracle, of which the wattled framework was covered with
hide; coracles are still in occasional use by fishermen on Welsh
rivers. Primitive peoples frequently make boats in this way.
Fig. 81 shows the framework of the Umiak, or women's boat
of the Eskimo, made of driftwood, laced together with thongs,

FIG. 82.—Eskimo Umiak.

Fig. 83.—A Sewn Bark Canoe.

without a single nail, and covered with skins; and Fig. 82 how it is fitted with a mast, and square sail of membrane. Fig. 83 is an interesting canoe made by the Australian natives, with bark sewn on to a framework. Fig. 84 shows swords of the Early Iron Age. No. 1 shows an early Halstatt pattern, and 2 a later La Tène type shown in scabbard. The scabbards were in bronze, and frequently ornamented with very beautiful designs. The sword blade was of iron, with a tang on to which was fitted a bronze mount to the handle, the latter formed of bone or wood threaded on to the tang.

Fig. 84 also shows two iron spear-heads of the same period which are rather different from the leaf-shaped patterns of

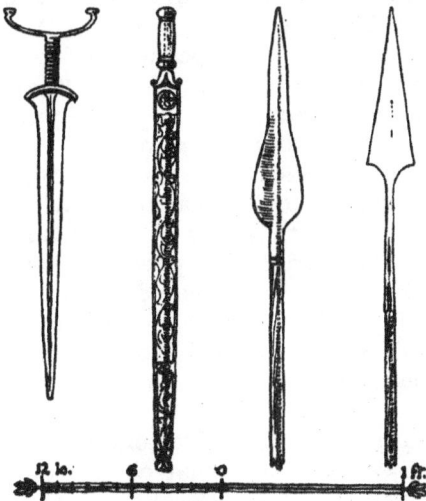

Fig. 84.—Early Iron Age Swords and Spears.

the Bronze Age. The shields were now oblong in shape, as that of the Belgic man in the costume plate (Fig. 56). This splendid work of art can be seen at the British Museum, and is made of bronze decorated with enamels. This form of

FIG. 85.—Enamelled Harness Ornament.

decoration appears to have developed out of the use of coral, added as an ornament to bronze. Then Early Iron Age metalworkers made studs, with an enamel surface, and pinned these to the bronze. This led the way to the crowning glory of his work, Champlevé enamelling. Here the field of the design was graved out of the metal, and the ground being first scored to give a key was filled in with the fused enamel, which, being polished, was finished flush with the face. Fig. 85, of an enamelled harness ornament, shows to what mastery of line the designers had now advanced. Think of the splendid appearance of an Early Iron Age chieftain ; his helmet, shield, and horse-mountings all bronze, not dull as now but shining like gold, with the enamels afire like liquid rubies. The earliest enamels were of one colour, red.

In the Early Iron Age, costume had developed and weaving in brilliant colours was practised. It is thought that these were combined into primitive tartans. As in the Bronze Age, a piece of material was folded around the body, in the form of a kilt, and this with a sleeveless vest, and a cloak which was semicircular in shape, completed a man's attire. The shoes were cut out of hide, with straps attached, and gathered round the ankle. The Brythons appear to have introduced the loose trousers, which originated with the Persians and Scythians. The women wore a long tunic reaching to the ankles, with short sleeves. Women, men, and horses, all alike, wore beautiful torcs, belts, and brooches, of bronze and enamel.

Another thing which was not found at Glastonbury was the burial-place, so that we do not know what objects they buried with their dead ; fortunately for archæologists, there are many other Early Iron Age cemeteries where this information can be gained. A very important one is at Arras, near

FIG. 86.—The Bronze Mirror.

Market Weighton, East Riding ; here the barrows are small, circular in form, not more than 2 feet high by about 8 feet diameter. The body was not cremated but buried in a very contracted position in a cist, or grave cut in the chalk. The skulls show the people to have been long-headed (*dolichocephalic*), and here for the first time iron is found with the body. This means either that there had been a reversion to the old burial customs of the Neolithic people, or these were introduced afresh from the Continent ; in any case the cremation of the Bronze Age passes away. Again, the long-headed skulls may point to a survival of Neolithic people, who had absorbed the old round-headed Bronze Age invaders, or to fresh invasions from the Continent. Some of the barrows at Arras, and in Yorkshire, were found to contain the remains of chariots, and these resemble the chariot burials in France ; this rather points to the Yorkshire barrows being the work of invaders. The tyres of the chariots there are about 2 feet 8 inches in diameter, and parts of the oak rims, or fellies were found, mortised for as many as 16 spokes. There were nave collars, for the hubs, of iron plated with bronze, and the skeletons of horses of about 13 hands. We saw the beginnings of chariots at Heathery Burn Cave in the Bronze Age, and it is obvious that by the time of the Early Iron Age these played an important part in everyday life. We have attempted a reconstruction in our Frontispiece, Fig. 1. Many of the Yorkshire barrows suggest that women were buried in them. In one was found one hundred glass beads of a beautiful deep blue colour, ringed and spotted with white ; others were of clear green glass with a white line. There were rings of amber and gold, and bracelets of bronze.

In the mounds were broken pottery, and the bones of animals, and charcoal, as if there had been a funeral feast. An iron mirror was found at Arras, very much rusted of course. Fig. 86 shows one of bronze of a more usual type.

We can now pass on to the latest type of burials in this country, and there is but little doubt that these were the work of Belgic invaders. They were discovered in 1886, at Aylesford in Kent. This was in the Belgic country, and here we find that cremation had again been introduced, and the Belgæ appear to have maintained this custom.

The cist, or grave, covered by a barrow, had passed out of fashion, and its place had been taken by a circular pit, about 3 feet 6 inches deep, the sides and bottom of which were daubed with chalky clay. In the pit were found burnt bones, and the fragments of the pottery cinerary urns, in which there had been placed a pail, flagon, skillet, or shallow saucepan, and brooches all of bronze. The custom evidently still persisted of burying objects which had belonged to the dead, because it had some symbolical meaning; or for their use in the spirit world; or because it would have been unlucky to retain the objects in everyday use. The pail is of the type carried by the Belgic girl in the costume plate (Fig. 56). The flagon of a very beautiful shape must have been imported from Italy.

The Aylesford pottery marks a great advance. It is of very graceful shape, and must have been turned on a wheel, and given a lustrous black surface in the firing. The wheel may have been of the turn-table type described on page 97, and shown at A in Fig. 87, or the potters may have advanced so far as the wheel shown at B. This is a very primitive type, which was used until lately for making flower-pots and bread-pans.

Except for this important detail of the reintroduction of cremation, the Belgæ do not seem to have effected any very great alteration in the everyday life of the times. They were a fierce fighting people, and conquered the S.E. districts. This gave them possession of the iron mines of the Sussex Weald, which was to be the Black Country of England until the eighteenth century.

The Brythons and the older Goidelic stock of the Bronze

FIG. 87.—A Potter's Wheel.

Age, and the people of Mediterranean descent as at Glastonbury, learned to use iron but continued to live their lives in their own way. Fig. 88 illustrates the use of bronze bowls as water-clocks. These were put to float in a larger bowl, and being perforated at the bottom, slowly filled, and in a certain time sank, and were then emptied by an attendant and refloated, to re-sink in another period. Fig. 89 shows late Celtic ornament. We saw by Fig. 61 how the Bronze Age peoples' patterns were chevrons, lozenges, and concentric circles, and the Early Iron Age saw the introduction of the curve, and the endless possibilities which come about through combination of curves.

There are still a great many things which remain a puzzle to the archæologist about the everyday doings of all those people who lived here before the coming of the Romans, when our written story began. Prominent among them is how they managed for communications and

FIG. 88.—Water Clock.

transport. It was seen (p. 44) how the stones forming the inner ring at Stonehenge, each weighing many tons, were brought all the way from Prescelly Mountains on the Pembrokeshire coast. This, as the crow flies, is about 150 miles. How could such a feat of transportation have been accomplished in 2000 B.C.? Either they were brought by land, making a wide détour to avoid the Severn Estuary, when thick virgin forest and wide marshes covered so much of the lower ground, or they must have come by sea, starting in the open Atlantic, weathering the rocky promontory of Cornwall, and beating up the chops of the English Channel. Yet we know nothing of their ships and very little of their roads, though a few of the many ancient trackways on the higher ground show some evidence of having been in use as early as the time of the Neolithic population. And why bring those great stones at all? Were they moved in slow stages as holy indispensibles in a tribal migration as the Ark of the Covenant was carried by the Israelites in the Bible? Or was it a mystical act of conquest such as prompted Edward the First to take the Stone of Scone from Scotland to London? Or was it that Prescelly Mountain was regarded as more sacred than Salisbury Plain and the transfer of its rock regarded as sanctifying the great shrine—just as we sometimes have water from the Jordan brought over here for christenings.

The tracks which are shown on the map (Fig. 3) are those which run along the crests of the chalk downlands and the

TRACKWAYS

Cotswold Hills. These all go over ground which has been less disturbed by the building and agricultural operations which have altered the face of this densely populated country during the Middle Ages and the more strenuous later times. For that reason they were used by the cattle-drovers right down to the end of the last century. So they have been well preserved and are well marked.

But the Bronze Age people were scattered all over the country, and the uniform even-ness of their culture indicates that they must have had a fair system of communications, if only by bridal-paths. Recent archæological research in Sussex has revealed the presence of Bronze Age farms (Fig. 60), and one observes that the roadways leading to these were fair-sized tracks and not mere bridle-paths.

As to the Iron Age it is most likely that there were paved roads. A settlement was excavated in Anglesey where a road of this kind and apparently of ancient date went right through its midst. And not far from the same place, in 1942, a most remarkable find was made of a quantity of iron tyres belonging to chariot-wheels. But if there were roads, even indifferent ones, there remains the problem as to who maintained them. That means a higher order of civilization than we usually credit our prehistoric ancestors with. Still, we must remember that the Bronze Age seems to have had more than a thousand years of peace, and if the tribes of the Iron Age were sufficiently well organised to make such vast earthworks as at Maiden Castle and Old Oswestry they may have maintained some sort of roads in the pauses between inter-tribal hostilities. Besides, in many things they imitated the Romans who were making such a stir with their conquests and their new way of things on the Continent. Already the Britons were imitating them in all sorts of ways. So that when the Cæsars arrived they had little difficulty in attracting the populations of the hill-forts into Romanised tribal centres.

Camulodunum, or Colchester, was the chief town of the Trinovantes ; Verulamium, or St. Albans, of the Catuvellauni, and Cassivellaunus was their king. Cæsar is supposed to have referred to St. Albans, when he wrote of "an oppidum with the Britons is a place amidst dense forest, fortified by a rampart or ditch, whither it is their habit to assemble to escape an enemy's raid." Corinium (Cirencester) was the home of the

FIG. 89.—Late Celtic Patterns.

CURRENCY

Calleva (Silchester) of the Atrebates ; London of the Cantii.
Women were allowed to be Queens. Cartismandua was
Queen of the Brigantes, and their country was the Pennines,
and Boudicca (Boadicea) of the Iceni.   •

In the Bronze Age chapter, we discussed Trade and Traffic
on page 76, and this brings up the question of money or the
currency which is used as a medium for that exchange of
goods, which is the basis of Trade.   It has been suggested
that the gold bracelets of the Bronze Age may have been used
as money ; these have been found with rings fastened to them,
and are called Ring-money, and the idea does not seem too
wildly remote.   This is hardly the case with Fig. 90, which
illustrates iron currency bars, and we can imagine our readers,
unless they are born financiers, saying, " How on earth could
anyone buy anything with a kind of iron walking-stick."   We
are quite sure that many boys and girls have been puzzled by
the various methods which have been adopted by different
peoples.   There was the British sovereign of gold, now
unhappily extinct ; its dirty greasy successor, so typical of
the time, the Treasury note ; one has heard of cowrie-shells,
and so on ; in all parts of the world different things seem to be
used, but none so odd perhaps as the iron bars of the Early
Iron Age.

Of the two currency bars found at Glastonbury, one is
27$\frac{7}{8}$ inches long, and weighs 4666 grains, the other, 21$\frac{1}{4}$ inches,
but much thicker than number one, weighs 9097 grains.
Mr. Reginald Smith has identified currency bars with the
taleæ ferreæ of Cæsar (*Bell. Gall.* v. 12), and it is thought
that there were six varieties, the British unit being about
4770 grains.   Bars of $\frac{1}{4}$, $\frac{1}{6}$, 1, 1$\frac{1}{2}$, 2, and 4 have been identified.
The map (Fig. 3) shows where bars have been found in Eng-
land, and is a proof of the wide distribution of trade even in
the Early Iron Age.

Perhaps we can give an illustration which will show how
these things become accepted as currency.   In remote
villages in this country not long ago, it was usual to have
a settling-up day once a year after harvest ; during the rest
of the year the people ran bills, which they chalked up on the
barn-door.   At settling time the farmer would go to the
miller and say, " How do we stand," to which the miller
replied, " I have ground your corn, but you had some of the
flour, and I sold the remainder, and owe you £5."   The miller

FIG. 90.—Currency Bars.

went to the baker who said, "Yes, I had my flour from you, but supplied you with bread, and owe you £5." The butcher bought his beasts from the farmer, but sold his meat to all the village, and so they weighed up the matter, and came to a settlement. It is quite conceivable that the same £5 note, with a little small change, would have passed from hand to hand, and enabled the village to start on another year's trading all square; if instead of the £5 note, you had an iron bar, it really did not matter so much—in fact it was rather better, because like our extinct gold sovereign, it was a thing of value itself, which is more than can be said of the Treasury note. Intertribal and international trade, though more complicated, was, and still is, conducted on this same basis, of the exchange of commodities. It is well to remember this, when so large a part of what is called business to-day is in reality only a gamble with the product of other men's labour. Real wealth springs from mother earth, and real work is to be engaged in winning or shaping her treasures.

We find a less extraordinary currency than the iron bars, about 150 to 200 B.C., in a British gold coinage of modern type of two values. This appears to have started in the S.E., and as some of these coins are inscribed, it shows that writing had progressed.

The unit system of the currency bars is proof of some system of weights and measures, and another is given by the beautiful pots, bowls, and metal work. A good craftsman does not make a thing to just any odd size. Use will have shown him what is the handiest weight, and the best size. A modern brick, for example, is of the size and weight that experience has shown the bricklayer can handle. Endless experiment has gone to prove this, and all the other details of everyday work and the tallies or the sticks, which were kept as a reminder, became in time recognized standards and measurements.

The currency bars are proof of the exchange of commodities, but do not help us to understand how values were fixed ; how much corn a plough was worth. With such necessaries of life, the plough was worth the extra amount of corn the farmer could grow by its use ; that would be its just price in theory. In practice it is often regulated by scarcity, which tends to increase the price of the plough, or by overproduction, when the price of ploughs goes down. Then there are luxuries, for which people will pay more than they are worth, because they are beautful, or very scarce, and so on. All this wants to be borne in mind ; we shall find how in the Middle Ages, Canon Law was very much concerned with the Just Price and Usury, and even to-day a Profiteer is not held to be a very pleasant person. Trade and currency bars ; weights and measures ; the honesty of the good man, and even the thieving of the rogue, are part of that wonderful peep-show into the past we call History, and cannot be neglected.

Now as we are approaching the end of our space, it may be as well to see if we can discover anything of the animating spirit which inspired these people, and gave savour to their everyday life. We saw in Neolithic times how men are thought to have worshipped the powers of Nature, with a great Mother God over all. Gildas, a monk, writing in the sixth century A.D., said : "Nor will I cry out upon the mountains, fountains, or hills, or upon the rivers, which now are subservient to the use of men, but once were an abomination and destruction to them, and to which the blind people paid divine honour." Yet Nature worship still lingers with stones which are lucky, and wells whose waters are curative.

Sun worship appears to have been typical of the early Bronze Age, and with the arrival of the Celts may have taken the form of Hero worship. It is probable that in the Early Iron Age, as the gods became more personal and intimate, they took to themselves as well the failings of man ; as they were stronger and braver than man, in the perpetual warfare they waged with the powers of darkness, so also they were more cruel and hard.

Druidism appears to have been the religion of the later Celtic tribes of Britain and Gaul, but doubtless it was grafted on to the Hero and Sun worship of the Bronze Age, and the older Nature and Moon worship of the Neolithic man. This has been a very general practice ; a conquering people would be willing to place the credit of the victory to the power of their own gods, yet unwilling to neglect the ones who had been overthrown. A god was a god, even when associated with defeat, and might easily revenge himself by alliance with the powers of Darkness. It was wiser then not to run any risks, so we find old Faiths adapted to New Religions.

Cæsar in *De Bello Gallico*, book vi., gives us an interesting picture of Druids and Druidism, and other sources of inspiration are the Celtic Myths and Legends that Mr. Squire has gathered together in his book. These tales have come down to us, because they were gathered together by monkish chroniclers, from the twelfth to the fifteenth centuries, but for all the time before that they had been traditional in the Celtic countries, since the days when they were first recited by Druidical bards to the accompaniment of harps.

Cæsar wrote of the Druids : " As one of their leading dogmas, they inculcate this : that souls are not annihilated, but pass after death from one body to another, and they hold that by this teaching men are much encouraged to valour, through disregarding the fear of death. They also discuss and impart to the young many things concerning the heavenly bodies and their movements, the size of the world and our earth, natural science, and of the influence and power of the immortal gods." Again quoting Cæsar : " The whole Gaulish nation is to a great degree devoted to superstitious rites ; and on this account those who are afflicted with severe diseases, or who are engaged in battles and dangers, either sacrifice human beings for victims, or vow that they will immolate themselves, these employ the Druids as ministers for such sacrifices, because they think that, unless the life of man be repaid for the life of man, the will of the immortal gods cannot be appeased. Others make wicker-work images of vast size, the limbs of which they fill with living men and set on fire."

From the little that is known, it can be gathered that the Druids formed a religious aristocracy, to which entrance could only be gained by a long novitiate. There was a Head,

CELTIC

or Pope, elected for life ; they were exempt from war and taxation ; acted as judges, and had a monopoly of learning. Time was reckoned by nights, and the year counted by the revolutions of the moon.   Fig. 88 shows a water-clock which is supposed to have been invented by the Druids.

White bulls were sacrificed before the mistletoe was cut from the sacred oak.   Captives were killed, and signs read from the flow of their blood, and the palpitation of their entrails.

The Gaulish Druids looked to their British brethren, as possessed of a purer faith, and novices were sent here to learn the mysteries.   This came about because the Continent fell under the influence of Rome at an earlier date than we did ; for the same reason, with the advent of the Romans here, Druidism was driven into the West, because its practices shocked even the Romans, until they finally routed it out of its headquarters in Anglesey.   It survived in Ireland, which never fell under the Roman influence, until S. Patrick overthrew Cromm Cruaich.

If the Celtic legends are poisoned by hints of awful cruelty, we must yet remember that it was not the cruelty of the Romans, who enjoyed the killing in the Amphitheatre, but the religion of sacrifice carried to its most awful conclusion. The Druids were not cruel for cruelty's sake, but to propitiate the gods.

On the other side of the picture, we have the pleasant fact that the Celtic Myths and Legends, becoming traditional, were handed down, and became in the hands of the monkish chroniclers the foundation on which has been built a Literature that is entirely our own.

We have seen what great artists the Celts were, when they turned to handicraft ; their metal work, and enamels, have been the inspiration of many an artistic revival, hailed as new, and yet in reality just as old as the Druids.

The great Celtic festivals were Beltane at the beginning of May, Midsummer Day, the Feast of Lugh in August, and Samhain.   We still have survivals of these in May Day, S. John's Day, Lammas, and Hallow-e'en or All Saints, and the bonfires around which we dance on joyful occasions, started life as the sacrificial pyres on which victims were burned to propitiate the gods, or cattle offered to stay the ravages of a murrain, or plague, at the original Celtic festivals.

There is a poem to Cromm Cruaich in the Books of Leinster which seems to us to explain the spirit of the times:

> " Here used to be
> A high idol with many fights,
> Which was named the Cromm Cruaich ;
> It made every tribe to be without peace.
>
> 'Twas a sad evil !
> Brave Gaels used to worship it.
> From it they would not without tribute ask
> To be satisfied as to their portion of the hard world.
>
> .   .   .   .
>
> To him without glory
> They would kill their piteous, wretched offspring
> With much wailing and peril,
> To pour their blood around Cromm Cruaich.
>
> Milk and corn
> They would ask from him speedily ,
> In return for one-third of their healthy issue :
> Great was the horror and the scare of him."

The books of Leinster were compiled early in the twelfth century by Finn macGorman, Bishop of Kildare, and he as a Christian may perhaps have twisted the tale a little to make Cromm a slightly worse fellow than he was, so as to emphasize the importance of his destruction by the " good Patrick of Macha," yet on the whole there is little doubt that in the days of the Druids, the world was ruled by Horror.

If we understand this, it also explains how it was, that when an obscure Jew was crucified in Palestine, and left behind a handful of disciples, who preached that God was Love, it came as a light to lighten the darkness in a world that was horrible to the poor and oppressed, and held little comfort for them ; and here, for a time, our story ends.

INDEX TO TEXT AND ILLUSTRATIONS

NOTE.—The ordinary figures denote references to pages of text, those in black type, the Illustrations.

H.G., used in the text, means *Everyday Things in Homeric Greece*, by the same authors.

A

Adze, 18, **94**
Agger, 28
Agriculture, 17, 27
Alignment, 41
Alpine, 9, **10**
Animals, 6
Arminghall, 50
Armlets, 58
Arrow-heads, **19**
Aryans, 12
Avebury, vii
Awls, 58
Axes and hammers, **17**, 18, 100
Azilian, 1

B

Badbury Rings, 28, 29, 31
Barrow, 71, 72, **73**
Beads, 106
Beaker people, 11
Belgæ, burials, 107
— coming of, 14
— dwellings, 84
Bellows, 93
Bermes, 30, 32
Bill-hook, **95**
Boring, 18
Bracelets, 98
Bread, 24
Bridge, **67**
Bridles, 58, 63, **65**
Bronze, 52
— Age barrows, 23
— Age men, 11, 51, 52
— Age smith, 57
Brooches, **58**, **95**, 96, **97**
Bryn Celli Dhu, 30, **31**, **32**
Brythons, vi, 14, 84
Bucket, 58
Bulb of percussion, 15, 16
Burial, 71, 106, 107
— methods of, 39
— mounds, 37

C

Camps, **5**, 6
Canoes, 2, **7**, 90, **104**
Carding, (wool), 59
Cattle, 6
Cave men, 1
Celtic art, 84
Celts, coming of, 12
— craftsmanship of, 66
— or axes, 17, 18, **51**, **53**, 55
Celtic patterns, 11
Chalk, 4
Chariots, 66, 92, 106
Chisel, 58
Cist, 71
Clapper Bridge, **67**
Clocks, **109**
Clothing, 23
Coal money, 98
Coldrum, 6
Comb, **63**
Cooking hearths, 21
Copper, 79
Coracle, **102**
Core, 15
Costume, **62**, 63, 105
Counterscarp, 30
Cromlech, 41
Crucibles, 93
Cultivation of land, 17
— evolution of, 27
Currency bars, **113**

D

Danish midden axe, **3**
Deadfall trap, **35**
Deer-horn implements, **14**
Dew pond, **33**
Dice, **100**
Discs, 58, 72
Dog, 3
Dolmen, 41, **42**
Domestication of Animals, **3**
Druids, 50, 115
Durrington, 50

INDEX TO TEXT AND ILLUSTRATIONS

E

Earth-house, **38**
Earthworks, 27
Elf darts, 20
Enamel (for ornament), 105
Escarpment, 30
Eskimo huts, **40**
European races, 8

F

Festivals, 116
Finger-ring, 95
Fires, methods of making, 24
Flakes, 15, 19
Flint making, 15
— implements, 14, **15, 16**
— miners, **13**
— spears, **19**
Fosse, 28

G

Gaels, 12
Gate into England, 4
Glastonbury, **83, 85**, 86, 87, **89, 91**
Goidels, vi, 11, 12, 14, 66, 82
Gold coinage, 113
Gouge, 58
Grimes' Graves, 80
Grinding corn, **24, 92**

H

Hafting, methods of, **16** 17, **51**
Hammers, 15, 17, 18, 28
Harness, development of, 64
Harvesting, 68, 69
Heathery Burn Cave, 58, 66, 106
Hector's funeral, 73
Hele Stone, 46
Herdsmen, 4
Heroes, 39
Hill forts, 7, 8, 28, 109
— construction of, 29
Homeric parallels, 18, 38, 39, 48, 52, 54, 57, 63, 64, 66, 69, 72, 74, 76
Houses (Neolithic), 20
Hubs, 58, 64
Hut circles, 20, 32
Huts, 69, **87**, 88
— development of, 22

I

Icknield Way, 6
Immortality, belief in, 39
Implements, 17
Iron, 93
— Age, 82

J

Javelins, 19

K

Kitchen middens, 2
Kitscoty, 6
Knives, 58, 94
Knowles experiment, 34

L

Lake Villages, 86
La Madeleine paintings, 49, 76
Lances, 19
Lathe, 66, 98, 99
Leverage, 43
Linces, the, Cheddington, Bucks, vi
Loess, 75
Long barrows, 23, 37
Loom, 61, 62
Lynchets, 17, 24

M

Maiden Castle, 6, 30, 32, 81
Map of England, xi, xii
Mediterranean men, 9
Megalithic builders, 44
— monuments, 10, 28, 41
Maen Hir, 41, 42
Metals, 51, 79
Migrations, 10, 12, 75
Mirror, 106
Money, 112
Monolith, 41, 42

N

Neolithic Age—Geographical conditions, 4
— cooking, 22
— hunters, 6
— huts, 21, 22
— implements, 14
— man, 7
— period, 8
Nordic, 9, 11
Nucleus, 15

INDEX TO TEXT AND ILLUSTRATIONS

O

Oppidum, 110
Ornament, 70, 105, 111

P

Palisades, 32
Pastures on Downs, 6
Pattern, 27
Peytrel, 65
Picks, 15, 30
Picts, 23
Picts House, 39
— tower, 41
Pilgrims' way, 6
Pins, 58
Pit dwelling, 20, 21
Plough, 68
Ploughing, 92
Postbridge, 67, 68
Potter's wheel, 108
Pottery, 25, 26, 27, 69, 71, 97, 107
Pounding grain, 25
Prickers, 58

Q

Querns, 24

R

Racial characteristics, 9
Rampart, 28
Rapier, 56
Razor, 58, 64
Revetment, 30
Ridgeway, 8
Rings, 58, 95
River drift, 1
Rivers, 7
Roman Camps, 28

S

Sacrifice, 48
Saw, 94
Scrapers, 18, 22
Shield, 58, 105
Ships, 78
Shovels, 15, 30
Sickle, 23, 24
Silbury Hill, 72
Skewers, 58
Slaves, 36, 39

Smelting, 22, 54, 93
Smith, 57
Social life, 35
Spears of bronze, 55
— of iron, 104
Spindle whorls, 58, 59
Spinning, 59, 60, 97
Standing Stone, 41, 42
Stonehenge, 42, 44, 45, 46, 47
— methods of construction, 44–45
Strike-a-light, 19, 22, 24
Sun temples, 48
Swiss lake dwellings, 84
Sword, 56, 104

T

Teasing (wool), 59
Terremare, 84
Tin, 79
Tongs, 58
Tools, 95
Trackways, 6, 8, 36, 80, 109
Trade routes, 77, 78
Traders and trading, 36, 76
Trilithon, 41
Trumpet, 68

U

Umiak, 103

V

Vallum, 28

W

War, 36
Warp, 60
Warriors and chariot of the Early Iron Age, *frontispiece*
Wattle and daub building, 90
Water clock, 108
Water supply, 32
— methods of obtaining, 32, 33
Weapons, bronze, 51
Weaving, 60
Weft, 61
Wheel, 64, 66
Woodhenge, 50
Wood turning, 97
Worship, 114